Intelligent Techniques for Data Science

Rajendra Akerkar • Priti Srinivas Sajja

Intelligent Techniques for Data Science

 Springer

Rajendra Akerkar
Western Norway Research Institute
Sogndal, Norway

Priti Srinivas Sajja
Department of Computer Science
Sardar Patel University
Vallabh Vidhyanagar, Gujarat, India

ISBN 978-3-319-80514-6   ISBN 978-3-319-29206-9   (eBook)
DOI 10.1007/978-3-319-29206-9

This Springer imprint is published by Springer Nature
The registered company is Springer International Publishing AG
The registered company address is: Gewerbestrasse 11, 6330 Cham, Switzerland

Dedicated to our parents

Preface

Information and communication technology (ICT) has become a common tool for doing any business. With the high applicability and support provided by ICT, many difficult tasks have been simplified. On the other hand, ICT has also become a key factor in creating challenges! Today, the amount of data collected across a broad variety of domains far exceeds our ability to reduce and analyse without the use of intelligent analysis techniques. There is much valuable information hidden in the accumulated (big) data. However, it is very difficult to obtain this information and insight. Therefore, a new generation of computational theories and tools to assist humans in extracting knowledge from data is indispensable. After all, why should the tools and techniques, which are smart and intelligent in nature, not be used to minimize human involvement and to effectively manage the large pool of data?

Computational intelligence skills, which embrace the family of neural networks, fuzzy systems, and evolutionary computing in addition to other fields within machine learning, are effective in identifying, visualizing, classifying, and analysing data to support business decisions. Developed theories of computational intelligence have been applied in many fields of engineering, data analysis, forecasting, healthcare, and others. This text brings these skills together to address data science problems.

The term 'data science' has recently emerged to specifically designate a new profession that is expected to make sense of the vast collections of big data. But making sense of data has a long history. Data science is a set of fundamental principles that support and guide the extraction of information and insight from data. Possibly the most closely related concept to data science is data mining— the actual extraction of knowledge from data via technologies that incorporate these principles. The key output of data science is data products. Data products can be anything from a list of recommendations to a dashboard, or any product that supports achieving a more informed decision. Analytics is at the core of data science. Analytics focuses on understanding data in terms of statistical models. It is concerned with the collection, analysis, and interpretation of data, as well as the effective organization, presentation, and communication of results relying on data.

This textbook has been designed to meet the needs of individuals wishing to pursue a research and development career in data science and computational intelligence.

Overview of the Book

We have taught the topics in this book in various appearances at different locations since 1994. In particular, this book is based on graduate lectures delivered by the authors over the past several years for a wide variety of data science-related courses at various universities and research organizations. The feedback from participants and colleagues at these venues has helped us to improve the text significantly.

The book can be used at the graduate or advanced undergraduate level as a textbook or major reference for courses such as intelligent control, computational science, applied artificial intelligence, and knowledge discovery in databases, among many others.

The book presents a sound foundation for the reader to design and implement data analytics solutions for real world applications in an intelligent manner. The content of the book is structured in nine chapters.

A brief description of the contents found within each chapter of the text follows.

- Data is a vital asset to any business. It can provide valuable insights into areas such as customer behaviour, market intelligence, and operational performance. Data scientists build intelligent systems to manage, interpret, understand, and derive key knowledge from this data. Chapter 1 offers an overview of such aspects of data science. Special emphasis is placed on helping the student determine how data science thinking is crucial in data-driven business.
- Data science projects differ from typical business intelligence projects. Chapter 2 presents an overview of data life cycle, data science project life cycle, and data analytics life cycle. This chapter also focuses on explaining a standard analytics landscape.
- Among the most common tasks performed by data scientists are prediction and machine learning. Machine learning focuses on data modelling and related methods and learning algorithms for data sciences. Chapter 3 details the methods and algorithms used by data scientists and analysts.
- Fuzzy sets can be used as a universal approximator, which is crucial for modelling unknown objects. If an operator can linguistically describe the type of action to be taken in a specific situation, then it is quite useful to model his control actions using data. Chapter 4 presents fundamental concepts of fuzzy logic and its practical use in data science.
- Chapter 5 introduces artificial neural networks—a computational intelligence technique modelled on the human brain. An important feature of these networks is their adaptive nature, where 'learning by example' replaces traditional 'programming' in problems solving. Another significant feature is the intrinsic

parallelism that allows speedy computations. The chapter gives a practical primer to neural networks and deep learning.

- Evolutionary computing is an innovative approach to optimization. One area of evolutionary computing—genetic algorithms—involves the use of algorithms for global optimization. Genetic algorithms are based on the mechanisms of natural selection and genetics. Chapter 6 describes evolutionary computing, in particular with regard to biological evolution and genetic algorithms, in a machine learning context.
- Metaheuristics are known to be robust methods for optimization when the problem is computationally difficult or merely too large. Although metaheuristics often do not result in an optimal solution, they may provide reasonable solutions within adequate computation times, e.g., by using stochastic mechanisms. Metaheuristics and data analytics share common ground in that they look for approximate results out of a potentially intractable search space, via incremental operations. Chapter 7 offers a brief exposure to the essentials of metaheuristic approaches such as adaptive memory methods and swarm intelligence. Further classification approaches such as case-based reasoning are also discussed in the chapter. This classification approach relies on the idea that a new situation can be well represented by the accumulated experience of previously solved problems. Case-based reasoning has been used in important real world applications.
- To achieve the benefit that big data holds, it is necessary to instil analytics ubiquitously and to exploit the value in data. This requires an infrastructure that can manage and process exploding volumes of structured and unstructured data—in motion as well as at rest—and that can safeguard data privacy and security. Chapter 8 presents broad-based coverage of big data-specific technologies and tools that support advanced analytics as well as issues of data privacy, ethics, and security.
- Finally, Chap. 9 gives a concise introduction to R. R programming language is elegant and flexible, and has a substantially expressive syntax designed around working with data. R also includes powerful graphics capabilities.

Lastly, the appendices provide a spectrum of popular tools for handling data science in practice. Throughout the book, real world case studies and exercises are given to highlight certain aspects of the material covered and to stimulate thought.

Intended Audience

This book is intended for individuals seeking to develop an understanding of data science from the perspective of the practicing data scientist, including:

- Graduate and undergraduate students looking to move into the world of data science.
- Managers of teams of business intelligence, analytics, and data professionals.
- Aspiring business and data analysts looking to add intelligent techniques to their skills.

Prerequisites

To fully appreciate the material in this book, we recommend the following prerequisites:

- An introduction to database systems, covering SQL and related programming systems.
- A sophomore-level course in data structures, algorithms, and discrete mathematics.

We would like to thank the students in our courses for their comments on the draft of the lecture notes. We also thank our families, friends, and colleagues who encouraged us in this endeavour. We acknowledge all the authors, researchers, and developers from whom we have acquired knowledge through their work. Finally, we must give thanks to the editorial team at Springer Verlag London, especially Helen Desmond, and the reviewers of this book in bringing the book together in an orderly manner.

We sincerely hope it meets the needs of our readers.

Sogndal, Norway   Rajendra Akerkar
Gujarat, India   Priti Srinivas Sajja
March 26, 2016

Contents

Chapter 1
Introduction to Data Science

1.1 Introduction

Data are raw observations from a domain of interest. They are a collection of facts such as numbers, words, measurements, or textual description of things. The word 'data' comes from '*datum*' and means 'thing given' in Latin. Data are ubiquitous and are important trivial units for instrumentation of a business. All entities directly or indirectly related to the business, such as customers of the business, components of the business and outside entities that deal with the business, generate a large pool of data. Data are often considered as facts, statistics and observations collected together for reference or analysis. Data provide the basis of reasoning and calculations.

Data can be qualitative as well as quantitative. Examples of qualitative data are people describing how luxurious a car is or the smell of a perfume (What a great smell . . . !). An example of quantitative data is a statement describing a car having four wheels (A car has four wheels). The last example, that of a car, describes countable items; hence, it is discrete in nature. On the other hand, a statement such as 'My weight is 150 pounds' is an example of continuous quantitative data. The height of a tree, the time taken in a race, and a person's height are also examples of continuous quantitative data.

It is believed that the word 'data' has been defined and used since beginning of the 1500s. With the advancements of computing technologies, the word became more popular. However, the word is not limited only to computer science or electronics fields, as applications in all fields use and generate data to some extent. In the current era, there has been an explosion of data. Tremendous amounts of data are generated from various resources every day. These resources include day-to-day transactions, data generated from sensors, data generated and stored on the Web and servers by surfing the web; and data created and provided by users. That is to say, data come from everywhere. Among these sources of data, the Web is the largest. Most people use the Web either as the basic infrastructure of their business, as a means of entertainment, and as a source to quench their thirst for the information. It

© Springer International Publishing Switzerland 2016
R. Akerkar, P.S. Sajja, *Intelligent Techniques for Data Science*,
DOI 10.1007/978-3-319-29206-9_1

has been observed that data expands to fill whatever storage space there is available. Most likely, you feel the desire to have more storage space! Storage capacity is easy to increase; however, a large amount of data increases the degree of complexity in managing it. If such data are made useful through proper techniques, then it could be a great help for problem solving and decision making.

Data Science is the systematic study and analysis of various data resources, understanding the meaning of data, and utilizing the data as an instrument for effective decision making and problem solving. Having knowledge from such data helps an organization to be efficient in terms of cost, delivery and productivity; identifies new opportunities, and creates a strong brand image of the organization. The purpose of data science is to facilitate applications of the various processes related to data such as data procurement, data pre-processing for noise clearing, data representation, data evaluation, data analysis and the usage of data to create knowledge related to the business. Data science contributes in terms of innovative methods to share, manage and analyse data in an optimized manner. The goal of data science is to discover knowledge that aids in decision making at individual, organizational and global levels. Besides identifying, collecting, representing, evaluating and applying data in order to discover knowledge, data science also facilitates data for effective utilization, and helps in optimizing data for cost, quality and accuracy. Conceivably, the ultimate emerging opportunity in data science is *big data*—the ability to analyse enormous data sets generated by web logs, sensor systems, and transaction data, in order to identify insights and derive new data products.

1.2   History of Data Science

According to John W. Tukey (1948), who coined the term Bit in 1947, data analysis along with the statistics it encompasses should be considered as an empirical science (Tukey 1962). Subsequently, Tukey (1977) also published his work entitled *'Exploratory Data Analysis'*, which emphasizes a more focused use of data to suggest hypotheses to test. He also pointed out that *Exploratory Data Analysis* and *Confirmatory Data Analysis* should proceed side by side.

A book written by Peter Naur (1974), a Danish scientist and winner of the ACM's A.M. Turing Award 1977, mentions a survey of contemporary data processing methods in the United States and Sweden. He defines data as 'a representation of facts or ideas in a formalized manner capable of being communicated or manipulated by some process'. He has used the words *'datalogy'* and *'science of data'* throughout the book. The roots of the book are in his presentation of a paper at the International Federation for Information Processing (IFIP) Congress in 1968, titled *'Datalogy, the science of data and of data processes and its place in education'* (Naur 1968). While providing the definition of field data science, he suggested that data science is the science of dealing with data and has relations with other fields. Many scientists such as Peter Naur dislike the term computer science and identify the field as *datalogy*.

In August 1989, Gregory Piatetsky-Shapiro organized a workshop on Knowledge Discovery in Databases (KDD). This was the first workshop in the history of data science and knowledge discovery. This started a series of KDD workshops that later grew into KDD conferences.

In September 1994, a cover story was published in Business Week[1] on database marketing, citing the large amount of information collected by companies about the customers and products. According to its claims, this information was used successfully to improve product sales. In March 1996, the term 'data science' was first used in the title of a conference as *'Data science, classification, and related methods'*. This conference was organized by the International Federation of Classification Societies (IFCS) at Kobe, Japan.

The term data science was also used by William S. Cleveland in the year 2001. He wrote *'Data Science: An Action Plan for Expanding the Technical Areas of the Field of Statistics'* (Cleveland 2001). In his paper, he has described an action plan to analyse data technically and statistically. He has identified six technical areas and establishes the role and importance of data science in these areas. This idea triggered the commencement of various events, pioneered by an international council for science: the Committee on Data for Science and Technology. The council also started a journal named *'The CODATA Data Science Journal'*[2] in April 2002. The next year, in January 2003, Columbia University also started a journal on data science, entitled *'The Journal of Data Science'*[3] Taking inspiration from these events, many researchers and institutes came forward to contribute in the field of data science. Data science soon emerged as a systematic study of data from various sources using scientific techniques. Some methods developed earlier as guidelines for processing and managing data also came into the limelight. Though the term *data science* was coined later in the twenty-first century, data processing methods were already in use.

Thomas H. Davenport, Don Cohen, and Al Jacobson published *'Competing on Analytics'* in May 2005.[4] They highlighted the use of statistical and quantitative analysis of data as a tool to perform fact-based decision making. Instead of considering the traditional factors, results from the aforementioned techniques are considered as primary elements of competition. The Harvard Business Review then published this article in January 2006,[5] and it was expanded (with Jeanne G. Harris) into a book with the title *'Competing on Analytics: The New Science of Winning'*[6] in March 2007.

In January 2009, a report entitled *'Harnessing the Power of Digital Data for Science and Society'* was published by a science committee of the National Science

---

[1]http://www.bloomberg.com/bw/stories/1994-09-04/database-marketing

[2]https://www.jstage.jst.go.jp/browse

[3]http://www.jds-online.com/v1-1

[4]http://www.babsonknowledge.org/analytics.pdf

[5]https://hbr.org/2006/01/competing-on-analytics/ar/1

[6]http://www.amazon.com/Competing-Analytics-New-Science-Winning/dp/1422103323

and Technology Council.[7] The report discusses the importance of a new discipline called data science and identifies new roles for the field. The report enlists roles such as digital curators (responsible for digital data collection), digital archivists (to acquire, to authenticate and to preserve the data in accessible form), and data scientists (responsible persons such as scientist, software engineers, domain experts and programmers for digital data management).

In May 2009, Mike Driscoll published '*The Three Sexy Skills of Data Geeks*', in which he established the importance of data scientists and identified such data geeks as a hot commodity.[8] According to him, skills such as the ability to statistically analyse data, mung data (cleaning, parsing, and proofing data before its use) and visualize data are essential skills in the field of data science.

In the year 2010, Mike Loukides in the article '*What is Data Science?*' identified the job of a data scientist as inherently interdisciplinary. He stated that such experts can handle all aspects of a problem, from initial data collection and data conditioning to the drawing of conclusions.[9] The interdisciplinary nature of data science is illustrated with the help of Venn diagrams having components such as mathematical and statistical knowledge, hacking skills and substantive expertise.[10]

1.3 Importance of Data Science in Modern Business

Though data are known as the currency of a new business era, it is not enough to just possess data. In order to attain better and effective utilization of the data available, data must be processed and analysed in proper ways to gain insights into a particular business. Especially when data is coming from multiple sources, in no particular format and with lots of noise, it must undergo processes of cleaning, munging, analysing and modelling. Data science has its applicability in all aspects of business. All business activity generates a lot of data. Having an abundance of such currency (data related to the business) should be a desirable situation; instead, such data create a large problem due to size, unstructuredness and redundancy. Some researchers identify parameters such as volume, velocity and variety as the main obstacles to handling data. According to Eric Horvitz and Tom Mitchell (2010) and James Manyika et al. (2011), such data, when analysed and used properly, offer a chance to solve problems, accelerates economic growth, and improves quality of life. It is really an irony that the same data that can help improve quality of life instead make our lives miserable because of our limitations to understand and use it properly. Many researchers and innovative contributors have provided useful models and techniques to deal with big data; however, a comprehensive and

---

[7]https://www.nitrd.gov/About/Harnessing_Power_Web.pdf

[8]http://medriscoll.com/post/4740157098/the-three-sexy-skills-of-data-geeks

[9]http://radar.oreilly.com/2010/06/what-is-data-science.html

[10]http://drewconway.com/zia/2013/3/26/the-data-science-venn-diagram

focused approach is needed. In a survey carried out by Thomas H. Davenport, Don Cohen and Al Jacobson (2005), some critical characteristics and parameters for data science practicing companies are identified. The survey includes about 32 companies that practice data science successfully in diverse disciplines and have gained competitive advantages on the basis of data science activities and analytics. According to the research, the main observations made for the companies are as follows.

- There is more than one type of data scientists as well as experts who practice analytics and data science-related activities for growth of the business.
- Not only statistics, but also in depth data analysis, modelling and visualization techniques are used for business-related decision making.
- The span of such data science activities are not limited to a small function of a business, but are applied to multiple business activities.
- Company strategy is inclined towards the use of analytical and data science activities.

Most companies are attracted to the application of data science to improve business; however, they do not know exactly how such activities can be planned or how business strategy should be modified. The first requirement is for skilful data scientists and experts who can envisage the possible organizational and technical benefits. It is necessary to envisage the requirement of resources and infrastructure to implement data science-related activities. It is also necessary to identify possible sources of data and permissions as well as the methods needed to acquire the data. Experts can also provide guidance about availability of other experts, tools and models that may be useful during the process. While foreseeing the possible activity map in advance, the experts may identify possible barriers. Once data scientists or experts are selected, their next step after such planning is to identify barriers towards the goal. The second step is learning and building data science skills. Statistical, modelling, programming, visualization, machine learning and data mining techniques are essential for doing data science. The third step is an action-oriented step. At the local level, identified barriers are removed and corrective actions are suggested. A major barrier to the application of data science is the availability of data, the collection of such data and the structuring of the gathered data to obtain sufficient meaning. A model appropriate to the collected data might be finalized. Here, an application-specific model or technique can be designed. The fourth step is implemented for data science activities using the collected data and selected designs. The collected data must be cleaned, analysed, and processed into a suitable model and presented in a good manner. In this phase, it can be possible to have minor variations in design to efficiently implement the model. These activities are illustrated in Fig. 1.1.

As this is an early version of data science activities, they are normally implemented at a local level or in a limited scenario. If the results are promising and in line with the business goals, then similar activities are designed and experimented at the organizational level in an extended form. Later, further applications of data science activities for competitive advantages are carried out in an integrated manner.

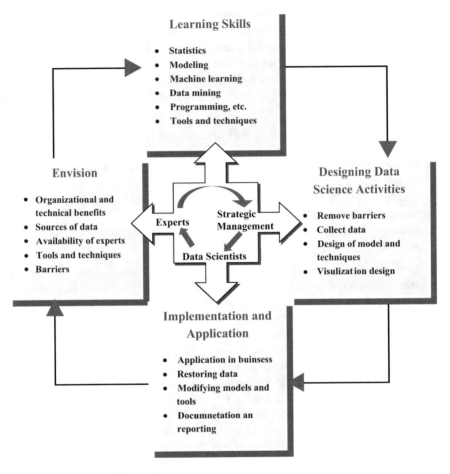

Fig. 1.1 Broad phases of data science

1.4   Data Scientists

A data scientist is the key person in acquiring, clearing, representing and analysing data for business and research purposes. He or she takes care of planning various business activities, coordinating between components of business and managing the life cycle of the business. To perform the job, a data scientist must have multi-domain and versatile skill sets. Above all else, a data scientist must have the ability to work on multiple projects simultaneously. Ideally, he or she is supposed to have a combination of skills in analytics, machine learning, and data mining and statistical data processing, and a little bit of computer programming is also desirable.

The data scientist's job will vary depending on an organization's nature, size and span. He or she may have to deal with a company whose main business activity is data handling. In this case, machine learning and statistical data processing are the important skills expected in a data scientist. Sometimes companies will, at some stage, require the handling of large pools of data through an efficient infrastructure. Here too, the data scientist is helpful in establishing data infrastructure and other resources including personnel. In this case, the data scientist is supposed to have an introductory background in software engineering. Some companies are not data driven or do not handle and analyse data, but have moderate amounts of data. Discoveries from such data might be highly useful to the company's business activities. Most typically, large companies fall in this category. In this case, besides the basic skills, a data scientist is meant to exhibit data visualization skills. Table 1.1 describes the basic skills essential to data scientists.

To be specific, a data scientist is a generalist in business analysis, statistics, and computer science, with a proficiency in fields such as robust architectures, design of experiments, algorithm complexity, dashboards, and data visualization, to name a few. A data scientist should serve as the champion of data science, with a mandate to combine internal and external data to generate insights that enhance a business's decision-making ability.

Table 1.1 Essential skills for a data scientist

Skill	Description	Applicability
Basic tools	Basic tools include office packages, charting tools, programming language such as R[a] or Python[b] and query language such as Structured Query Language (SQL).	All types of organizations.
Basic statistical knowledge	Statistical models, statistical tests, distributions, and estimators.	All type of organizations, especially applicable to product driven companies where sufficient amount of data are to be handled to make effective decisions.
Machine learning	Techniques such as k-nearest neighbours, random forests, ensemble methods, classification	Most suitable for data driven organizations where data is the main product. Machine learning skills are helpful in automatic analysis or smart analysis of the data.
Calculus and linear algebra	Many of these skills are actually from basis of different machine learning techniques. Understanding of such techniques allows the data scientist to modify the techniques in innovative manner. A small improvement by such innovative implementations can result in multi-fold improvement.	Used when large amount of data related to a product, users or business is to be managed.

(continued)

Table 1.2 (continued)

Skill	Description	Applicability
Data munging	Dealing with unformatted, partial, incomplete and ambiguous data. Techniques to properly format data, finding and predicting missing data and identifying meaning of ambiguous data are useful in cleaning and preparing data for further processing and use.	Data driven companies dealing with large amount of data.
Data visualization and communication	Effective graphing tools, data description and visualization tools such as Dygraphs[c]. It is a good way to communicate with non-technical users.	Most suitable for the data driven organization where data are used to support crucial decisions.
Software engineering	Techniques related to requirements collection, portfolio management, personnel and other resources management, and data logging techniques are helpful.	Such techniques are useful in data driven organization when developing products and services that deal with lots of data.

[a]http://www.r-project.org/
[b]https://www.python.org/
[c]http://dygraphs.com/

1.5   Data Science Activities in Three Dimensions

To take complete advantage of data science, a data scientist should consider many things simultaneously. There are three main dimensions to a data scientist's activities. They are, namely: (1) data flow, (2) data curation and (3) data analytics. Each dimension focuses on a category of problems and methods of solutions for the problems. It is the data scientist's responsibility to perform investigations and studies in all three dimensions simultaneously, and to propose a holistic plan for data used for decision making and problem solving. Figure 1.2 briefly outlines these activities with respect to the aforementioned dimensions.

1.5.1   Managing Data Flow

The first dimension of data science activities is managing *data flow*. Logically, this activity comes first. It starts with the *collection of data*. Using fact-finding methods such as interview, questionnaire, record review and observation, one can find the required data. With the advancement of Information and Communication Technology (ICT), such as Internet, the Web and other modern devices, information can be collected efficiently. Time period and frequency of data collection are

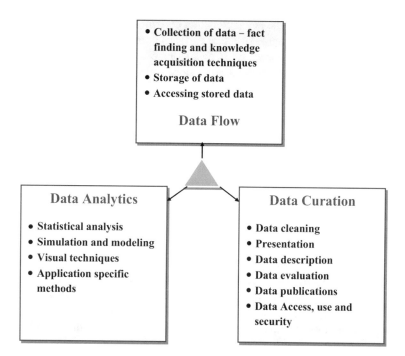

Fig. 1.2 Data science activities in three dimensions

parameters that also have to be considered by data scientists in order to efficiently collect data. 'When to collect data' and 'how to collect data' play important roles in collecting meaningful and useful data. Prior to data collection, a design of data collection activity has to be ready. This activity includes listing possible resources of data, types of data to be collected, variables for which data are required, encoding data for better representation, personnel to be involved in data collecting process, and actual data collection. Collected data can be in multi-media form. It can be textual information, photocopies of existing documents and records, audio/video recordings of interviews, and graphs/charts of procedures. The data scientist also has to deal with the problems of managing devices and technology, such as the format of audio files and images and accidental loss of data while collecting. Sometimes transcribing (converting other multi-media format into textual information) also needed to be carried out by data scientists. Coding of data is very helpful when dealing with qualitative data. Qualitative data encoded in proper format are comparatively easy to analyse. It is also useful to arrange the collected data in a proper manner for future use.

Another activity under the dimension of the data flow is *storage of data*. Collected data must be stored for its future use. Often, data are collected and organized in such a way as to fit into the predefined storage structure. This should be the parallel procedure. Storage structure of the data can be decided in parallel with the data collection format. Data types, as well as their nature, form and frequency of utilization, are some of the key factors that determine data and storage structure. It is better to decide the storage structure early, as immediately after collection, the data must be preserved somewhere. If data storage is not designed or finalized, a temporary storage structure can be used to store collected data. Later, the data can be stored in suitable form.

The stored data must be accessible, useful, transparent enough, in its possible native form, and complete to some extent. Further, the data scientist may store metadata (for easy access) and a replica of the data (for backup and experimentation), depending on the nature of application and requirement of the business. To accommodate a high volume of data, a distributed approach is used; where data can be divided and stored at different locations using a format supported by local infrastructure. That is, data can be distributed among various platforms and hardware according to its planned use. Data scientist may opt for cloud-based infrastructure as an alternative platform to store collected data, to efficiently manage capacity and accessibility of the data. However, putting sensitive data on the cloud also raises security and network issues. Access efficiency of data on the cloud also depends on bandwidth and other network infrastructure. It must be noted that the main characteristic of the data being handled in this context is that it is big in nature. Therefore, it is very important that the selected storage structure must not be costly. As a high amount of memory space is needed to accommodate the collected data, cost-effective storage structure is the prime requirement of the business.

Accessing the Stored Data in an efficient way is the third activity in data flow management. The accessing mechanism depends on the nature of the data collected and the storage medium chosen. Flat files, relational databases along with query language support, and XML-based representation are popular mechanisms of data access. These methods are very efficient as far as structured data are concern. When it comes to a large amount of unstructured data, the traditional methods of data accessing performance are not appreciable. To cover this limitation, an innovative idea of relaxing the constraints on the methods was proposed. Researchers have tried to relax data availability and consistency, and have proposed Not-only Structured Query Language (NoSQL) technologies. As its name denotes, the term specifies that the databases do not use only Structured Query Language (SQL) to access the data. The idea is to restrict and simplify fully functional database technologies. Here data scientists can play important roles. They can determine how the traditional database constraints should be relaxed, whether row data should be partitioned horizontally or vertically, what the access patterns are and how the workflow is to be managed. Data scientists are also responsible for documenting performance and operational characteristics of the data. Figure 1.3 illustrates major contributing techniques to managing the flow of data.

In addition to the collection, storage and access of data, there also needs to be a sufficient level of data mobility. As data volume, velocity and unstructuredness increase, mobility of the data on a given network decreases. Increasing the capacity of the network (such as 100 gigabits per second or even more) is not always practically feasible. Many times, it is necessary to process such data at the source

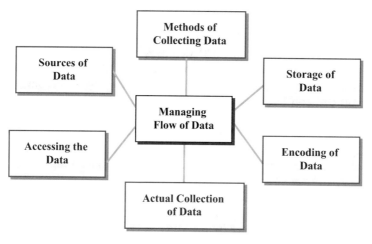

Fig. 1.3 Managing flow of data

itself to reduce the cost of data movement. Sometimes, this introduces many errors in data or data loss. Data scientist may think about data processing and compressing techniques before the data really goes mobile.

1.5.2 Managing Data Curation

Data curation activities deal with manipulating collected data in a sophisticated way. The word curation is also used in the humanities and museum fields, where information and articles are kept for information and exhibition. It is obvious that once collected, the data cannot be used in their original form. The collected data must be cleared from noise. It is also necessary to check that the data are complete or if some portion is missing. If so, one has to determine whether the missing data can be ignored or not and if it is necessary to search for the missing data. The data can be recollected or some smart technique can be applied to fill the gap with imaginary (dummy) data. If data are imprecise, they should be converted into a form with the required precession before any further analysis. Generally, data are in native form. Here also, it is necessary to convert the data into a uniformly acceptable (and if possible, electronic) format. It is better if such formatted data are machine consumable and can be reused by human beings as well as machines in future. This is actually a continuous approach, starting from the collection of data, sophisticated storage and usage of the data until the data become obsolete and goes to the bin. Major activities to be named under data curation are conceptualization of data, creating missing data, clearing data, representing data for efficient accessing and use, data description, evaluating data for an application, data preservation and security, reusing of data, documentation and disposal of the data. Among these,

(1) data preservation, (2) data presentation and description, (3) data publication and (4) data security are the major activities.

Data Preservation depends on the way data are stored. Many times, while collecting data from old documents (such as stone carvings, old papers, cloths and leaves on which data is carved), the data are lost or damaged. The curator's prime focus is on collecting as much as complete and sufficient data from the article. Data are preserved in various formats and mediums. Each medium has its own pros and cons. Digital data storage medium, for example, can preserve data more efficiently, but it requires the data to be digitalized first. Even if the data are stored in secondary memory such as hard disks, this is for a limited time span. Typical hard drives can keep your data safe for 3–6 years. Flash drives keep data safe for 80–90 years. A flash drive can safely preserve the data for approximately a century because it does not use magnetic field to store data; this makes such a drive magnet-proof and thus safer and longer lasting. Type of data, bulk of data, frequency of use, cost of the storage medium and nature of the domain from which the data has come are important parameters in choosing data preservation methods. Depending on requirements, the data can also be compressed and encrypted before preserving the data; this makes the bundle of collected data smaller or bigger in comparison with the native data collected from the source.

Data Presentation and Description activities help to place the data in a sophisticated manner. The nature of data is that it does not convey any meaning. The data must be processed to obtain some understanding out of it. The data must accompany something that can explain something about the data. Data description defines the use and meaning of data. Providing a description about data particularly helps in understanding that data. Data structures, schemas, main variables, aliases, metadata and useful content related to data (such as standards used to describe data) are the main components used in the description of the data. With such components, it is easy to find purpose, meaning and use of the data. For some new generation applications (such as web based applications), data ontology is also considered. If data are described properly, chances of better utilization of data are increased.

Data Publication activities help to make data available to the target audience. Prior to publication, the collected data must be cleaned, formatted and described well for its effective use. For data stored in the digital media in a suitable format, publication of the data is faster and cheaper. Automatic use as well as transfer of the data also becomes efficient. Searching from such a data source is easier too. In cases where the data are available in forms other than digital media, automatic scanners and programs are available to convert the content into a digital equivalent. On the other hand, this increases the chances of misuse of the data if proper security measures are not adopted. The privacy of sensitive data must be maintained.

Many formats are used while digitalizing the data; these include photocopies; video or audio files, or simply text entries of the data to be published. Another important technique is to store data with some links. Such interlinked data are useful for hiding unnecessary information at a given stage, and provide a platform

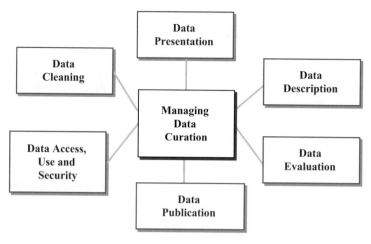

Fig. 1.4 Data curation

for semantic queries on the data. Digitalization of data can be done manually or automatically, such as with 'heads-up' technology, where a good quality image (preferably in raster format) of the data is projected for better visualization by users. Users view such images with their 'heads up' and digitalize the data. Sometimes the help of technology is used to read the images automatically. Figure 1.4 illustrates major contributing techniques of data curation.

Data security activities specifically focus on making data secure from accidental loss or manipulations. Such activities are critical for most businesses, especially when data are the backbone of the business and kept at central locations to facilitate simultaneous access. Sensitive data, such as clients' data, payment-related records, policy information about business and personal files, must be preserved from falling into the wrong hands. If such data are lost, comparatively less damage is done; but if such data fall into malicious hands, this can create bigger problems. Hence, it has become necessary to identify possible risks and threats to the data. Some of the risks are as follows.

- Physical threats – such as infrastructural damage, fire, flood, earthquake, and theft.
- Human error – data are misinterpreted and/or misused by humans; data is accidentally deleted and dropped data while modifying structures; and input and/or processing errors. One may lose or damage data while converting the data into digital format.
- Planned exploitation of data through malicious activities.

The data are normally available to the collector, curator, administrator and high-level users for various purposes. Data may be vulnerable via any of them. It is the data scientist's responsibility to employ suitable methods to secure the data and make them available to the users on demand. Many methods of securing data are available. Some of the popular methods to secure digitalized data are as follows.

- Protection of data and system with alarms and monitoring systems.
- Implementing a firewall and keeping anti-malware system up-to-date.
- Keeping an eye on hacking attacks with the help of intrusion detection technology and modifying anti-malware systems accordingly.
- Checking status of operating system. The operating system must be accurate to deal with data smoothly with efficient memory management and resource allocations.
- Auditing systems and intelligent encryption mechanisms.
- Infrastructural safety and alarming system against theft and physical damage.

Security measures also depend on the data's nature, bulk, complexity and the storage medium used to store the data. If, for example, mobile applications are considered, security means should also be applied on the mobile devices. Further, at regular intervals, the data must be backed up. Though several methods are available to make data more secure, some additional effort must be taken to develop innovative application-specific methods for data security.

Safe and well-managed data are a good and authentic resource to generate new data through transformation. Data curation, if done properly, adds great value to the data collected.

1.5.3 Data Analytics

After successful acquisition and pre-processing of data, the data must be analysed in order to find meaningful information and patterns. Data analytics procedures transform data into meaningful summaries, reports, patterns and related visual information that provides useful information to support decision-making processes in a business. They generate new ideas and innovative mechanisms in business procedures. Proper utilization of analytics techniques on the available data help to describe, predict, and improve business performance. Analytics techniques are most appropriate when business data are large and unstructured. To uncover hidden patterns in the data, analytics techniques such as simulations and modelling, statistical analysis, machine learning methods and methods for visual analytics are generally used. The approach for applying analytics can be reactive or proactive. In the case of a reactive approach, once data are available, selected analytics methods provide reports and insights based on data. In the case of a proactive approach, data analytics methods are invoked first and continuously search for the appropriate data, using a list of potential resources. In a proactive approach, only those data that have some potential in terms of analytics are extracted.

Analytics based on *statistical analysis* focus on application of statistical models on the data acquired. Popular methods are predictive models, matrix valued analysis and prediction models. Some of the models are very effective on standard-size data sets; however, when it comes to large data sets, the typical models are not successful. Special models, such as principal component analysis, compressive sampling, clustering, need to be developed by doing research in data science.

Typical *modelling and simulation* techniques are also to be reinvented. The traditional theories and model use predefined concepts such as boundary, target, entity, attributes, status and constraints. For a small set of data, many simulation and modelling techniques work, but for a big set of data, such techniques are not very useful. As the data resources are disorganized, it is difficult for typical models to provide fruitful solutions. For any modelling and simulation practice, there needs to be a good set of data. Here, a lot of data is available, meeting the prime requirement. It is a challenge to researchers and developers to identify and develop new models and simulations specially meant for large and unstructured data.

Visual Models and Techniques help in presenting data (or interpretations derived from the data) in a visual manner for easy understanding and communication to users. When data are large and in diverse forms, it is a challenge to represent them visually. Special Visual Analysis (VA) tools should be used for visual analysis of large sets of heterogeneous data.

Research is also emerging in the fields of Web, mobile, sensor and network data; referred as Web analytics, mobile analytics, sensor analytics and network analytics, respectively. These are subfields of data and text analytics. In these fields, techniques developed for data analytics are useful with minor changes. Figure 1.5 illustrates major contributing techniques of data analytics.

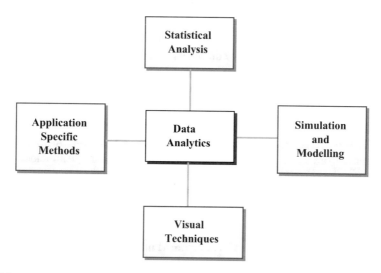

Fig. 1.5 Data analytics

1.6   Overlapping of Data Science with Other Fields

The first field that comes to mind is *Information and Communication Technology (ICT)*. The Internet as service-providing platform with necessary techniques and protocols; and the World Wide Web (Web), as a kernel lying over the platform of the Internet, serves as means of sharing data. It is the ICT that made the platform available to generate, share and use the data, as well as the methods and techniques that efficiently use the source. The field of ICT also provides support in the form of client server and other distributed architectures, data communication methods, data encryption, decryption and compression techniques, programming languages, data plumbing, beside the International Organization For Standardization (ISO)'s basic Internet model and protocols.

Another field closely related to data science is *statistics*. Data are the backbone of the field of statistics. Data and statistics share a tight relationship, as do either side of a coin. That's why data are often misinterpreted as statistics. Commonly, the words data and statistic are used interchangeably. However, data are raw observations from the domain, whereas statistics refer to statistical tables, charts, graphs and percentages generated from the data using well-defined methods and models. Raw data are a source for statistics. Raw data can also be in the form of organized database files, text file or any machine-readable file. If such digitalized data are available, statistical software can be used to generate statistics from the data. The field of statistics deals with organizing data, modelling data, and applying techniques such as multi-variate testing, validation, stochastic processes, sampling, model-free confidence intervals, and so on.

Machine Learning is considered as an integral component of computer science and a field related to the ICT. Machine learning is a technique of data analysis that mechanizes analytical model building. Using algorithms that iteratively learn from data, machine learning lets computers to discover hidden insights without being explicitly programmed where to look. Machine learning is a field used when searching the Web, placing advertisements, stock trading, credit scoring and for several other applications.

Data Mining is closely related to machine learning and data science. This discipline contributes algorithms and models that help in extracting meaningful data and patterns from large data sets. Some of the typical techniques include pattern recognition, classification, partitioning and clustering along with a few statistical models. In other words, data mining has some overlap with statistics too. As ordinary data have expanded, edits span to multi-media data, and data mining is also stretched to multi-media mining.

Mining of images, and videos have emerged into context, and today researches concede to obtain knowledge from these types of data in a semantic manner. It is evident that it is becoming difficult to find a scenario where data mining is not used.

Operations Research also uses data, fitting them into appropriate models and providing cost–benefit analysis to support decisions. The major aim of the operation research is application of a suitable model to support the decision-making process. Operation research aids business activities such as inventory management, supply chain, pricing and transportation, by using models such as linear programming, Markov Chain models, Monte-Carlo simulations, queuing and graph theory.

Business Intelligence is also a field related to data science. However, the main goal of business intelligence is to manage information for better business. The field of business intelligence deals with techniques and models that generate information useful for analysis, reporting, performance management, optimizing decisions, and information delivery.

Artificial Intelligence emphasizes the creation of intelligent machines that work and react like humans. Examples include: surveillance systems, self-driving cars, smart cameras, robotic manufacturing, machine translations, Internet searches, and product recommendations. Modern artificial intelligence often includes self-learning systems that are trained on huge amounts of data, and/or interacting intelligent agents that perform distributed reasoning and computation. Artificial intelligence is the field that is ubiquitously applied in most other fields and can contribute in any domain. Artificial intelligence has the ability to learn from vast amounts of data, and the power of simulating bio-inspired behaviour besides typical intelligent models and algorithm. This makes artificial intelligence universally applicable where a typical formal model fails.

Besides the aforementioned fields and areas, data science experts also need to be master other domains such as communication, entrepreneurship and art and design (to present data in visual manner). Figure 1.6 shows the relation of other fields to data science.

1.7 Data Analytic Thinking

One of the most vital facets of data science is to sustain data-analytic thinking. The skill of data analytical thinking is key not just for the data scientist, but for the entire organization. For instance, managers and line employees in other functional areas can only obtain the best from the organization's data-science resources if they have understanding of the basic principles. Managers in organizations without significant data science resources should still understand basic principles in order to involve consultants on an informed basis. Venture capitalists in data science need to know the basic principles in order to assess investment opportunities. Knowing the key concepts, and having frameworks for organizing data-analytic thinking, will not only allow one to interact proficiently, but it will help in foreseeing prospects for refining data-driven decision making, or predicting data-oriented competitive risks.

Nowadays, companies are using data analytics as a competitive differentiator. As a result, a new business function is being formed within all companies: a data analytics team. The primary goal of this team is to help leaders view business problems from a data perspective and to bring in data analytic thinking as an input into the decision-making process.

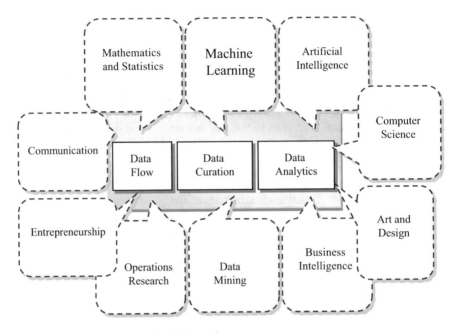

Fig. 1.6 Relation of other fields with data science

1.8 Domains of Application

There are many domains that make use of data. Government and administration, manufacturing industries, service industries, private sectors, and research and development related domains are some typical domains that generate and use tremendous amounts of data. In the following subsections, we discuss some such instances.

1.8.1 Sustainable Development for Resources

From government offices, administrative units, non-profit and non-government organizations and from public resources, tremendous amounts of data about sustainable development are available and shared through satellite and via the Web and other media. These include data on infrastructure, population, natural resources, livestock and agricultural activities, population and climate. Understanding such data in a holistic manner provides new ideas and innovations for better development of an area. It is critical to analyse such multi-dimensional data from different angles, to make optimum use of the resources. Transportation facilities, irrigation, agricultural activities including marketing of crops, generation of traditional electricity, solar and renewable energies, and management of hospitals, schools, and other administration facilities should ideally be linked together. Managing such diverse data at once is a challenge because of the data's volume, diverse nature and format, and availability at one common place. A coordinated effort is required to collect data from various sources, analyse them, link them and build a model for development, which is a very difficult task. Practically, such a task as a single solution is highly ambitious and nearly impossible. However, the goal can be divided into many subprojects, and the help of data science activities can be used to achieve them in a phase-wise form. For example, one may start with sustainable water resource management. Besides existing and authentic data from government offices, data can also be taken from sensors, satellites, remote servers and the Web. Meaningful information, patterns and models can be derived from such data and presented through a data visualization process. Such data is useful to hydrologists, engineers, and planners, by providing hidden knowledge stored in the form of patterns. Once the goal is achieved, on the basis of the project outputs, a model for renewable energy or irrigation activities can be designed. Once all such phases are successfully designed, an integrated solution can be planned. Due to the nature of the problem, and uncertainty in the field and climate continuously changing a user's requirements and administration policies, it is difficult to achieve 100 % success in such a project, but a working solution is always provided with the help of data science activities.

1.8.2 Utilization of Social Network Platform for Various Activities

Many of us use social media platform to be in touch with other professionals, friends and colleagues. Young people especially survive on information and feedback from social media platforms. There is much professional and personal data as well as traces of activities available on such platforms, which is not utilized for any meaningful purpose. Such data can be used to promote a genuine product, increase awareness about selected systems or services, and promote learning in a special way. Healthcare, education, marketing, employment, identification of security threats, and social awareness programs can be accelerated using this already established media. The challenge is that most of this data is personal in nature; one should have proper permission and access to the data set generated by such platforms. Further, these data are totally unstructured, redundant and lack standard format. Beyond these difficulties, such data are full of errors and emotions. Even if the challenge of having accessibility to such data is met, another big challenge is to make sense from such data and establish a generic data-based model to promote the intended task. Once these two challenges are solved, such data are useful for a number of applications, including promoting government schemes, getting feedback on product, service and employees (evaluation and promotion), establishing e-learning, mobile learning and personalized learning projects on such platforms, generating innovative ideas for development, and troubleshooting and online assistance.

1.8.3   Intelligent Web Applications

The Web is an ever-increasing repository of data generated by many activities throughout the globe. Earlier, operations on the Web were restricted. The Web was considered as a 'read-only' type of resource. Gradually, the Web evolved as a 'read-write' platform. Today, one can upload, download and execute user defined contents and programs on the Web. It has evolved to a kind of 'read-write-execute' interface. There is a need for an effective search engines that searches according to users' context, manages native languages, saves time and offers customized results as per the original search goals. For example, while searching, we often jump to sites and lose the original goal of searching. Further, a user might type in 'tables and chairs', meaning 'mainly tables, but chairs will also do'. Boolean operators used in searches give standard weightage (in the case of AND operator, equal weightage is given) to both the keywords; this is not very suitable to the user's query. Actually, in this case, the user wanted more tables and fewer chairs. Such fuzzy models can also be applied while searching a large pool of data such as the Web (Sajja and Akerkar 2012). The volume, complexity, unstructuredness and continuously evolving nature of the Web provide enough challenges to users and applications that explore the Web. Here also, data science helps in acquiring data, analysing data and preparing visual outputs for better understanding and problem solving. Doing this, the Web can best be utilized for decision making using existing resources of tremendous amounts of data.

1.8.4   Google's Automatic Statistician Project

Even though vast amounts of data are available, because of aforementioned hurdles, it is difficult to use them in problem solving. Google also believes that it is difficult to make sense from such large pools of data. According to Google,[11] interpreting such data requires a high level of intelligence. In this situation, to expect a machine to collect, analyse, model and use such data from resources such as World Wide Web is considered to be highly ambitious. To achieve this ambitious goal, Google has started a project called Automatic Statistician. The aim of the project is to traverse all available resources of data, identify useful data, and apply suitable statistical methods to discover a good understanding of the data in order to produce a detailed report in native language. Some example analyses done at the early stages of the project are provided on the site.[12] The project uses various algorithms on provided data sets and generates detailed reports with necessary figures. For example, data about an airline demonstrating non-stationary periodicity are provided to the project. The project uses the Automatic Bayesian Covariance Discovery (ABCD) algorithm

---

[11] http://www.automaticstatistician.com/
[12] http://www.automaticstatistician.com/examples.php

and produces a detailed report including executive summary, identifies four additive components from the data, and discusses them in detail with necessary tables and figures. Other examples are interpretation of data related to solar irradiance, sulphuric acid production data, and unemployment-related data analysis. The data regarding these examples are analysed using the aforementioned ABCD algorithm.

1.9 Application of Computational Intelligence to Manage Data Science Activities

Most of the data science-related activities have to manage large pools of data in many ways. These data have some common characteristics that make handling of such data very difficult. Some of the characteristics are given below.

- *Volume of data*: Any activity leaves its traces in the form of some data. Especially, when a business transaction is performed online using communication technology, it generates digital data. Such data and data about the data (metadata) are generated without any control; rather, it is believed important to have more data to be on safer side. However, just generating and posing such data will not help to improve business, but it creates the additional problem of storing and maintaining the data for future use. Further, the data in their native form are not very useful. They must be cleaned, analysed and modelled before being used for decision making. The volume of such data creates the biggest hurdle. Volume-associated problems can be efficiently handled by applying smart and intelligent techniques. An example of efficient handling of such large volumes of data is provided by Google. A few years ago, Google announced a remarkable finding in the journal Nature. According to the claim, Google's research team was quick enough to be first to track the spread of influenza across the US. They did it without consulting a physician and without performing a single medical check-up. Such speedy predication is done by tracking the outbreak by finding a correlation between what people searched for online and whether they had flu symptoms.[13] Such an efficient, timely and cost-effective solution is possible because of ability of the Google to handle such large volumes of data and ability to set non-traditional, out-of-box type hypotheses.
- *Variety of data*: Data occur in many types, structures and formats. All of them are necessary and required to be stored and managed. Because of advancements in modern technology, it has become easier to store photos, videos and audio files beside textual information. Such variety of media and structure is another major hurdle in handling data acquisition, storage, analysis and applications. As stated earlier, data are never self-explanatory. They are hard to characterize and difficult to process, especially when collected from diverse resources and forms.

---

[13] http://www.ft.com/cms/s/2/21a6e7d8-b479-11e3-a09a-00144feabdc0.html#axzz3SqQt4zUS

- *Veracity (quality of data)*: Often data contain biases, noise and abnormality within them. Some noise is introduced at the acquisition phase and some is inserted while working with data, such as while formatting, storing and backing up. Sometimes interpretation and processing of the correct data can also be done in an incorrect way, resulting in infeasible outcomes. Irrespective of the volume of data, such veracity may exist due to aforementioned reasons. It is necessary to consciously try to clean dirty data by employing various techniques to separate unwanted data from that which is useful.
- *Validity of data*: It is possible that the data are correct, but not appropriate and suitable. It is not that noise is introduced into the data, but that the data are not valid for deducing any conclusion. Some real-time data will not be useful after a given time. It is also necessary to determine how long to store the data and when a data set loses its validity. If it is known, unwanted data can be moved to some parallel storage schemes.
- *Data are redundant and contradictory*: Data are ubiquitous and generated simultaneously by many parallel activities at multiple locations. It is quite possible that some data are stored in duplicate, using other formats or in modified version. Such redundancy of the data is necessary to identify. It is desirable to identify such duplicate data and convert the data into a standard uniform format, so that comparison and identification of duplicate data is easy.
- *Data are partial and ambiguous*: As we know, data do not convey any meaning. Unprocessed data cannot provide meaningful information. As mentioned above, data themselves are not self-explanatory and can be interpreted in many ways. Sometimes, data are lost during the process of acquisition and communication. Before interpreting and analysing the data, lost data must be compensated. This can be done by finding the lost data from the original source, or creating dummy data to compensate for the loss. Ambiguous and partial data can be filled and repaired by artificial intelligent techniques such as fuzzy logic. Figure 1.7 illustrates the characteristics that make data handling difficult.

According to the well-known data pyramid, data is the key ingredient to generate information. Many chucks of the data are processed systematically in order to produce information. Information is not always directly available from the sources. For example, the average turnover of a company is not available, but has to be deduced from the individual data regarding turnover of the company every year. Such multi-dimensional information is synthesized and know-how type of knowledge is generated. When such knowledge becomes mature, wisdom is generated. Wisdom is often defined as a possession and use of diverse knowledge on its best requirement. It is wisdom that ensures the possessed knowledge is used in such a way that goals and objectives of one's life are not violate, but their achievement is facilitated. The journey of collecting and processing data to get information and model, to be able to know how and learn a better use of knowledge ends at being intelligent. This shows that data are basic and important raw materials to generate intelligent behaviour. Such tremendous amounts of data are continuously generated and available via many platforms, including the Internet and Web. If such

Fig. 1.7 Characteristics that make data handling difficult

data are used properly, it will be a great help towards achieving highly effective and intelligent decision making within a business. This leads to application of computational intelligence techniques for better utilization of available data.

1.10 Scenarios for Data Science in Business

As we have discussed in earlier sections, data science is becoming an integral part of in-house functionalities of an organization, just as Accounting, Legal aspects and information Technology (IT) are. Currently, some big data start-ups are being absorbed by large IT vendors who develop big data *enterprise applications* and convince corporate buyers that big data analytics is the next big thing. The bottom line is that an enterprise is inundated with data and the potential to tap into that data to capture market share is an opportunity organizations cannot afford to squander.

Here are some standard scenarios for data science in the field of business:

1. Greater levels of understanding and targeting customers: A large retailer has been able to accurately predict when a customer of theirs is expecting a baby. Churn management has become easily predictable for communication firms, and car insurance companies are able to understand how well their customers are driving.

2. Smarter financial trading: High frequency trading is finding huge application for big data now. Data science algorithms used to make trading decisions has led to a majority of equity trading data algorithms, taking data feeds from social media networks and news websites into account to make split second decisions.
3. Optimizing business process: Data science is also an important method for introspection into business processes. Stock optimization in retail through predictive analysis from social media, web trends and weather forecasts is indicating vast cost benefits. Supply chain management is remarkably gaining from data analytics. Geographic positioning and radio frequency identification sensors can now track goods or delivery vehicles and optimize routes by integrating live traffic data.
4. Driving smarter machines and devices: The recently launched Google's self-driven car is majorly using big data tools. The energy sector is also taking advantage by optimizing energy grids using data from smart meters. Big data tools are also being used to improve the performance of computers and data warehouses.

These are some of the current examples where data science or big data is in application in the business sector. There are many other openings where data science can drive organizations into being smarter, secured and connected.

1.11 Tools and Techniques Helpful for Doing Data Science

For different phases of data science activities, supporting tools and techniques are available. This section describes tools and techniques for various activities such as data acquisition, data cleaning, data munging, modelling, simulation and data visualization.

Techniques such as structured interview, unstructured interview, open-ended questionnaire, closed-ended questionnaire, record reviews and observation are collectively known as fact-finding methods. Such fact-finding techniques and other data acquisition techniques can be automated instead of taking manual approaches. Usage of physical devices such as terminals, sensors and scanner with dedicated software is also available to manage the interface between the physical devices and system. Later, such data can be partly managed through typical programming languages such as Java, Visual Basic, C++, MatLab,[14] and Lisp. Open source and specialized data acquisition software such as MIDAS (Maximum Integration Data Acquisition System)[15] is also available. Often, data acquisition systems are developed as a dedicated, stand-alone system. Such systems are called data loggers. In the case of special need, a working model of a system is prepared and

---

[14]http://in.mathworks.com/products/matlab/

[15]https://midas.triumf.ca/MidasWiki/index.php/Main_Page

presented to the data scientists. Such a prototype is helpful for users to test the acquisition mechanism before it is actually built. It helps in collecting additional requirements and tests the feasibilities of the proposed system. There are knowledge acquisition and machine learning methods that find higher level content (such as information and knowledge automatically from resources). Examples of such knowledge acquisition methods are concept mapping, auditing, neural network and other methods related to automatic knowledge discovery. Among other tools, data cleaning tools, data munging and modelling tools, and data visualization tools are important. Some prominent tools in different categories are listed in this section. A more comprehensive list of tools is provided in Appendices I and II.

1.11.1 Data Cleaning Tools

Once collected, data needs to be checked for its cleanliness. Data cleaning, often known as data scrubbing, is the process of removing or correcting unwanted data from the source. The objective of the data claiming procedure is to identify and remove errors in data to provide consistent data for further analysis, modelling and visualization. At the entry level itself, some incorrect data are rejected through proper validations. In a homogeneous collection of data such as files and databases, the degree of inconsistency and amount of errors are less. However, in large databases with heterogeneous nature from multiple data sources, such as data warehouses, federated database systems or global web-based systems, the data cleaning becomes essential. Such problems occur due to (1) different formats, (2) redundant data, (3) difference in terminology and standards used for data and (4) methods of consolidations used for the data. Removing inaccurate, incomplete, or unreasonable data increases the quality of the data. Missing values, special values, range checks, deductive corrections, imputations, minimum value adjustments, typos, auditing and workflow specification checking are popular mechanisms for data cleaning.

Besides programming languages, the modern tools for data cleaning are listed as follows.

- Lavastorm Analytics[16] for products such as analytical engines.
- IBM InfoSphere Information Server[17] analyses, understands, cleanses, monitors, transforms and delivers data.
- SAS Data Quality Server[18] cleanses data and executes jobs and services on the DataFlux Data Management Server.

---

[16] www.lavastorm.com

[17] http://www-03.ibm.com/software/products/en/infosphere-information-server/

[18] www.sas.com

- Oracle's Master Data Management (MDM)[19] is a solution to handle voluminous data and to provide services such as to consolidate, cleanse, enrich, and synchronize key business data objects in the organizations.
- Experian QAS Clean service[20] provides CASS certification (Coding Accuracy Support System) for address verification services.
- NetProspex[21] supports data cleansing, appending and on-going marketing data management. It is now part of Dun and Bradstreet Information Services Pvt. Ltd.[22] in India, which offers data management transition and data quality programs.
- Equifax[23] provides database solutions for products for database management, data integration, and data analytics solutions.
- CCR Data cleanses and audits data. This company is developer of ADAM – the data cleansing platform.
- Solutions provided by the Oceanosinc organization[24] for data cleansing, contact discovery and business intelligence.
- Tools by Nneolaki[25] for data collection, cleansing, appending and management.
- Data cleanser products[26] for data cleansing solutions.

1.11.2   Data Munging and Modelling Tools

Other important activities while practicing data science are data munging and preparation. This is also known as data wrangling. It is a process of converting or mapping data into a well-formatted stream of data, so that data can be smoothly used for further processing. It actually allows convenient and automatic (partially or fully) consumption of the data with tools for further activities. Sorting, parsing, extracting, decomposing, and restoring data are the major activities under the data munging phase. Programming tools such Pearl, R, python, and some readymade libraries from programming languages and packages can be used to support data munging activities.

Once data are ready to analyse, statistical modelling techniques such as linear regression, operations research methods and decision support systems are typically used for data modelling. Here, the basic aim of the data modelling is to identify relations between the clean and valid data entities, in order to grow business insights.

---

[19] http://www.oracle.com/partners/en/most-popular-resources/059010.html

[20] http://www.qas.co.uk/

[21] http://www.netprospex.com/

[22] http://www.dnb.co.in/

[23] http://www.equifax.co.in/

[24] http://www.oceanosinc.com/

[25] http://neolaki.net/

[26] http://www.datacleanser.co.uk/

Data scientists or experts dedicated to this phase are known as data modellers. Data modelling can be done at conceptual level, enterprise level, and physical level. The following are the major tools that support data modelling.

- CA ERwin Data Modeler[27] data modelling solution provides a simple, visual interface to manage complex data.
- Database Workbench[28] offers a single development environment for developing with multiple databases.
- DeZign for Databases[29] is a tool that supports database design and modelling. It also offers a sophisticated visual data modelling environment for database application development.
- Enterprise Architect[30] is a fully integrated graphical support tool for data modelling and software engineering.
- ER/Studio[31] supports collaborative mechanism for data management professionals to build and maintain enterprise-scale data models and metadata repositories.
- InfoSphere Data Architect (Rational Data Architect)[32] is a collaborative data design solution. It simplifies warehouse design, dimensional modelling and change management tasks.
- ModelRight[33] is a tool for database designer to support activities such as database design, graphical support, reports and innovative visual interfaces.
- MySQL Workbench[34] is a unified visual tool for database architects, developers, and DBAs. MySQL Workbench provides data modelling, SQL development, and comprehensive administration.
- Navicat Data Modeler[35] helps in creating high-quality logical and physical data models.
- Open ModelSphere[36] is a platform-independent and free modelling tool available as open source software. It provides general support to all phases of data modelling and software development.
- Oracle SQL Developer Data Modeler[37] is a free graphical tool used to create, browse and edit data models. It supports logical, relational, physical, multi-dimensional, and data type models.

---

[27]http://erwin.com/products/data-modeler

[28]http://www.upscene.com/database_workbench/

[29]http://www.datanamic.com/dezign/

[30]http://www.sparxsystems.com/products/ea/

[31]http://www.embarcadero.com/products/er-studio

[32]http://www-03.ibm.com/software/products/en/ibminfodataarch/

[33]http://www.modelright.com/products.asp

[34]http://www.mysql.com/products/workbench/

[35]http://www.navicat.com/products/navicat-data-modeler

[36]http://www.modelsphere.org/

[37]http://www.oracle.com/technetwork/developer-tools/datamodeler/overview/index.html

- PowerDesigner[38] manages design time changes and metadata.
- Software Ideas Modeler[39] supports modelling through standard diagrams such as UML, Business Process Model and Notations (BPMN), System Modelling Language (SysML), and many diagrams.
- SQLyog[40] is a powerful MySQL manager and administrative tool.
- Toad Data Modeler[41] is a database design tool to design new structures as well as entity relationship diagrams along with an SQL script generator.

1.11.3 Data Visualization Tools

Data visualization refers to the graphical representation of data. Visualization of data makes it easy to understand and communicate.

There are many tools available for visualization of data. A brief list is given as follows.

- Dygraphs[42] is a quick and flexible open-source JavaScript charting library that allows users to explore and interpret dense data sets. It is a highly customizable tool.
- ZingChart[43] is a JavaScript charting library that provides quick and interactive graphs for voluminous data.
- InstantAtlas[44] offers interactive map and reporting software in an effective visual manner.
- Timeline[45] makes beautiful interactive timelines.
- Exhibit[46] is developed by MIT as fully open-source software. It helps to create interactive maps, and other data-based visualizations.
- Modest Maps[47] is a free library for designers and developers who want to use interactive maps.
- Leaflet[48] is a modern open-source JavaScript library for mobile-friendly interactive maps.

---

[38] http://www.powerdesigner.de/

[39] https://www.softwareideas.net/

[40] https://www.webyog.com/

[41] http://www.toad-data-modeler.com/

[42] http://dygraphs.com/

[43] http://www.zingchart.com/

[44] http://www.instantatlas.com/

[45] http://www.simile-widgets.org/timeline/

[46] http://www.simile-widgets.org/exhibit/

[47] http://modestmaps.com/

[48] http://leafletjs.com/

- Visual.ly[49] helps in creating visual representations.
- Visualize Free[50] creates interactive visualizations to illustrate the data beyond simple charts.
- Many Eyes by IBM[51] helps you to create visualizations from data sets and enables analysis on data.
- D3.js[52] is a JavaScript library that uses HTML, SVG, and CSS to generate diagrams and charts from multiple resources.
- Google Charts[53] presents a mechanism to visualize data in form of various interactive charts such as line charts and complex hierarchical tree maps.
- Crossfilter[54] is a JavaScript library for exploring large multi-variate datasets in the browser. It also offers co-ordinated visualizations in 3D.
- Polymaps[55] provides quick presentation multi-zoom data sets over maps.
- Gephi[56], as per the claim of developers, is an interactive visualization and exploration platform for all kinds of networks and complex systems, dynamic and hierarchical graphs. It supports exploratory data analysis, link analysis, social network analysis, and biological network analysis. This tool presents coloured regions for identified clusters of similar data.

Besides the aforementioned tools and techniques, many other dedicated and innovative tools are required in the field of data science. Since the field of data science is a consortium of techniques from many disciplines and has ubiquitous applications, it must be given prime importance in research and development. Also, there is a need of documentation, innovative techniques and models in the field of data science. As stated in this chapter, typical models and techniques may not be suitable for computing with the acquired data set, which may require support beyond their reach. Here, artificial intelligence techniques may contribute significantly.

1.12 Exercises

1. Explain the difference between data science and business intelligence.
2. Give examples of data analytics practice by businesses. How many of these enterprises achieved a competitive edge in the marketplace through harnessing the power of computational intelligence tools and techniques?

---

[49] http://create.visual.ly/

[50] http://visualizefree.com/index.jsp

[51] http://www-969.ibm.com/software/analytics/manyeyes/

[52] http://d3js.org/

[53] https://developers.google.com/chart/interactive/docs/

[54] http://square.github.io/crossfilter/

[55] http://polymaps.org/

[56] https://gephi.github.io/

3. Prepare a case study to demonstrate the application of the computational intelligence approaches in an industry sector of your choice.
4. Elaborate your understanding of the term 'data-driven decision making' with an example.
5. List as many fields as you can whose aim is to study intelligent behaviour of some kind. For each field, discover where the behaviour is manifest and what tools are used to study it. Be as moderate as you can as to what defines intelligent behaviour.

References

Cleveland, W. (2001). Data science: An action plan for expanding the technical areas of the field of statistics. *International Statistical Review, 69*(1), 21–26.

Davenport, T., Cohen, D., & Jacobson, A. (2005, May). *Competing on analytics.* Retrieved March 4, 2015, from www.babsonknowledge.org/analytics.pdf

Horvitz, E., & Mitchell, T. (2010). *From data to knowledge to action: A global enabler for the 21st century.* A computing community consortium white paper.

KDD. (1989, August 20). Retrieved March 4, 2015, from http://www.kdnuggets.com/meetings/kdd89/index.html

Manyika, J., Chui, M., Brad, B., Bughin, J., Dobbs, R., Roxburgh, C. et al. (2011, May). *Big data: The next frontier for innovation, competition, and productivity.* Retrieved March 4, 2015, from http://www.mckinsey.com/insights/business_technology/big_data_the_next_frontier_for_innovation

Naur, P. (1968). *Datalogy, the science of data and data processes and its place in education* (pp. 1383–1387). Edinburgh: IFIP Congress.

Naur, P. (1974). *Concise survey of computer methods.* Lund: Studentlitteratur.

Sajja, P. S., & Akerkar, R. A. (2012). *Intelligent technologies for web applications.* Boca Raton: CRC Press.

Tukey, J. (1948). A mathematical theory of communications. *The Bell System Technical Journal, 27*, 379–423.

Tukey, J. (1962). The future of data analysis. *Annals of Mathematical Statistics, 33*, 1–67.

Tukey, J. (1977). *Exploratory data analysis.* Reading: Pearson.

Chapter 2
Data Analytics

2.1 Introduction

In this digital era, data is proliferating at an unprecedented rate. Data sources such as historical customer information, customer's online clickstreams, channel data, credit card usage, customer relationship management (CRM) data, and huge amounts of social media data are available. In today's world, the basic challenge is in managing the complexity in data sources, types and the velocity with which it is growing. Obviously, data-intensive computing is coming into the world that aims to provide the tools we need to handle the large-scale data problems. The recent big data revolution is not in the volume explosion of data, but in the capability of actually doing something with the data; making more sense out of it. In order to build a capability that can achieve beneficial data targets, enterprises need to understand the data lifecycle and challenges at different stages.

Nowadays, there is a debate on data focused on the technology aspect. However, much more than technology is required to set up the fundamental basis of managing data analysis. It does not regard discarding existing structures, warehouses and analytics. Instead, it needs to build up on existing capabilities in data quality, master data management and data protection frameworks. Data management needs to be seen from a business perspective, by prioritizing business needs and taking realistic actions.

Data are basically of four categories: structured, semi-structured, quasi-structured and unstructured.

Structured data is that which is available in a pre-set format, such as row- and column-based databases. These are easy to enter, store and analyse. This type of data is mostly actual and transactional.

Unstructured data, on the other hand, is free form, attitudinal and behavioural, and does not come in traditional formats. It is heterogeneous, variable and comes in multiple formats, such as text, document, image, video and so on. Unstructured data

© Springer International Publishing Switzerland 2016
R. Akerkar, P.S. Sajja, *Intelligent Techniques for Data Science*,
DOI 10.1007/978-3-319-29206-9_2

is growing at a super-fast speed. However, from a business benefit viewpoint, real value and insights reside in this huge volume of unstructured data, which is rather difficult to control and channelize.

Semi-structured data lies somewhere between the structured and unstructured types. It is not organized in a complex manner that makes sophisticated access and analysis possible; however, it may have information associated with it, such as metadata tagging, which allows contained elements to be addressed.

Quasi-structured data is text data with inconsistent data formats.

Businesses have been inhibited in their ability to mine and analyse the huge amounts of information residing in text and documents. Conventional data environments were designed to maintain and process structured data—numbers and variables—not texts and pictures. A growing number of businesses is now focusing on integrating this *unstructured* data, for purposes ranging from customer sentiment analysis to analysis of regulatory documents to insurance claim adjudication. The ability to integrate unstructured data is broadening old-fashioned analytics to combine quantitative metrics with qualitative content.

Data always have a source. Just as big as data are, so are there diverse sources that can produce up to one million terabytes of raw data every day. This magnitude and dispersal in data is not of much use, unless it is filtered and compressed on the basis of several criteria. The foremost challenge in this aspect is to define these criteria for filters, so as to not lose out on any valuable information. For instance, customer preference data can be sourced from the information they share on major social media channels. But, how should we tap the non-social media users? They might also be a valuable customer segment.

Data reduction is a science that needs substantial research to establish an intelligent process for bringing raw data down to a user-friendly size without missing out the minute information pieces of relevance, when it is in real-time, as it would be an expensive and difficult issue to store the data first and reduce later. An important part of building a robust data warehousing platform is the consolidation of data across various sources to create a good repository of master data, to help in providing consistent information across the organization.

Data that has been collected, even after filtering, is not in a format ready for analysis. It is has multiple modes of content, such as text, pictures, videos, multiple sources of data with different file formats. This requires a sound data extraction strategy that integrates data from diverse enterprise information repositories and transforms it into a consumable format.

Once the proper mechanism of creating a data repository is established, then begins fairly complex procedure of data analysis. Data analytics is one of the most crucial aspects, and there is room for development in the data-driven industry. Data analysis is not only about locating, identifying, understanding, and presenting data. Industries demand large-scale analysis that is entirely automated, which requires processing of different data structures and semantics in a clear and computer intelligent format.

Technological advancements in this direction are making this kind of analytics of unstructured data possible and cost effective. A distributed grid of computing resources utilizing easily scalable architecture, processing framework and non-relational, parallel-relational databases is redefining data management and governance. In the big data era, database shave shifted to non-relational to meet the complexity of unstructured data. NoSQL database solutions are capable of working without fixed table schemas, avoid join operations, and scale horizontally.

The most important aspect of success in data analytics is the presentation of analysed data in a user-friendly, re-usable and intelligible format. The complexity of data is adding to the complexity of its presentation as well. Occasionally, simple tabular representations may not be sufficient to represent data in certain cases, requiring further explanations, historical incidences, and so on. Moreover, predictive or statistical analysis from the data is also expected from the analytics tool to support decision making.

Finally, the zenith of the entire data exercise is data interpretation or data visualization. Visualization of data is a key component of business intelligence. Interactive data visualization is the important thing the industry is moving into. From static graphs and spreadsheets to using mobile devices and interacting with data in real time, the future of data interpretation is becoming more agile and responsive.

In this chapter, we describe different lifecycles, from data to data analytics, and explore related aspects.

2.2 Cross Industry Standard Process

CRoss Industry Standard Process for Data Mining (CRISP-DM) offers constructive input to structure analytics problems (Shearer 2000). The CRISP-DM model has traditionally defined six steps in the data mining life cycle, as illustrated in Fig. 2.1. Data science is similar to data mining in several aspects; hence, there's some similarity to these steps.

The CRISP model steps are:

1. Business understanding – This initial phase focuses on understanding the project objectives and requirements from a business perspective, then converting this knowledge into a data mining problem definition and a preliminary plan designed to achieve the objectives.
2. Data understanding – The data understanding phase starts with an initial data collection and proceeds with activities in order to become familiar with the data, to identify data quality problems, to discover first insights into the data, or to detect interesting subsets to form hypotheses for hidden information.
3. Data preparation – The data preparation phase covers all activities to construct the final data set from the initial raw data.
4. Modelling – In this phase, various modelling techniques are selected and applied, and their parameters are calibrated to optimal values.

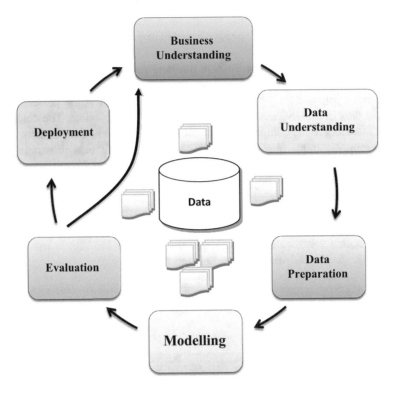

Fig. 2.1 Cross industry standard process

5. Evaluation – At this stage, the model (or models) obtained are more thoroughly evaluated and the steps executed to construct the model are reviewed to be certain it properly achieves the business objectives.
6. Deployment – Creation of the model is generally not the end of the project. Even if the purpose of the model is to increase knowledge of the data, the knowledge gained needs to be organized and presented in a way that the customer can use it.

CRISP-DM is particularly complete and documented. All the stages are duly organized, structured and defined, allowing a project could be easily understood or revised.

2.3   Data Analytics Life Cycle

Data analytics is a broader term and includes data analysis as a necessary subcomponent. Analytics states the science behind the analysis. The science is the understanding the cognitive processes an analyst uses to understand problems and explore data in meaningful ways. Analytics also include data extract, transform, and load; specific tools, techniques, and methods; and how to successfully communicate results.

In other words, data analysts usually perform data migration and visualization roles that focus on *describing the past*; while data scientists typically perform roles manipulating data and creating models to *improve the future*. Analytics is used to describe statistical and mathematical data analysis which clusters, segments, scores and predicts what scenarios are most likely to happen. The potential of analytics to determine the likelihood of future events is basically possible through tools such as online analytical processing (OLAP), data mining, forecasting, and data modelling. The process comprises analysing current and historical data patterns to determine future ones, while prescriptive capabilities can analyse future scenarios and present the most viable option for dealing with them. Analytics is at the core of data science and plays a vital role in its simplification, both during the initial phase of testing unstructured data and while essentially building applications to profit from the knowledge such data harvests.

However, analytics is not only about technology, hardware and data. It requires a cultural change in thinking. Thus the support for analytics cannot be only IT driven. It has to have business ownership if it is to succeed.

In the recent big data revolution, data analytics is gaining importance and is expected to create customer value and competitive advantage for the business. Major steps in the data analytics life cycle are shown in Fig. 2.2:

1. *Business objective*
   An analysis begins with a business objective or problem statement. Once an overall business problem is defined, the problem is converted into an analytical problem.

2. *Designing data requirement*
   To perform the data analytics for a specific problem, data sets from related domains are needed. Based on the domain and problem specification, the data source can be determined, and based on the problem definition, the data attributes of these data sets can be defined.

3. *Pre-processing data*
   Pre-processing is used to perform data operation to translate data into a fixed data format before providing data to algorithms or tools. We do not use the same data sources, data attributes, data tools, and algorithms all the time, as all of them do not use data in the same format. This leads to the performance of data operations, such as data cleansing, data aggregation, data augmentation, data sorting, and data formatting, to provide the data in a supported format to all the data tools as well as algorithms that will be used in the data analytics. In case of big data, the data sets need to be formatted and uploaded to Hadoop Distributed File System (HDFS) and used further by various nodes with Mappers and Reducers in Hadoop clusters.

4. *Performing analytics over data*
   Once data is available in the requisite format for data analytics, data analytics operations will be performed. The data analytics operations are performed to

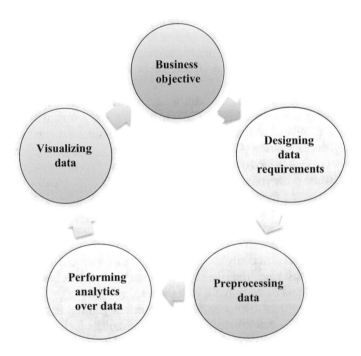

Fig. 2.2 Data analytics life cycle

discover meaningful information from data to make better business decisions. Analytics can be performed with machine learning as well as intelligent algorithmic concepts. For big data, these algorithms can be converted to MapReduce algorithms for running them on Hadoop clusters by translating their data analytics logic to the MapReduce job, which is to be run over Hadoop clusters. These models need to be further evaluated as well as improved by various evaluation stages of machine learning concepts. Improved or optimized algorithms can provide better insights.

5. *Visualizing data*

   Data visualization is used for displaying the output of data analytics. Visualization is an interactive way to represent the data insights. This can be done with various data visualization software and utilities. We will discuss more aspects of visualization in the last section of this chapter.

2.4 Data Science Project Life Cycle

In many enterprises, data science is a young discipline; hence, data scientists are likely to have inadequate business domain expertise and need to be paired with business people and those with expertise in understanding the data. This helps data

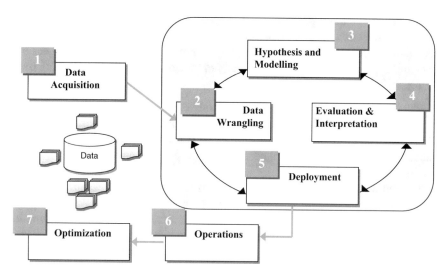

Fig. 2.3 Data science project life cycle

scientists work together on steps 1 and 2 of the CRISM-DM model—i.e., business understanding and data understanding.

The standard data science project then becomes an engineering workflow in terms of a defined structure of phases and exit criteria, which allow the making of informed decisions on whether to continue projects based on pre-defined criteria, to optimize resource utilization and maximize benefits from the data science project. This also prevents the project from degrading into money-pits due to pursuing non-viable hypotheses and ideas.

The data science project life cycle, proposed by Maloy Manna (Manna 2014), is modification of CRISP-DM with an engineering focus, as shown in Fig. 2.3.

Data Acquisition comprises obtaining data from both internal and external sources, including social media or web scraping. Data may be generated by devices, experiments, sensors, or supercomputer simulations.

Data wrangling includes cleaning the data and reshaping it into a readily usable form for performing data science. This phase covers various activities to construct the final data set (data that will be fed into the modeling tools) from the initial raw data. This activity can be performed multiple times, and not in any prescribed order. Tasks also include table, record, and attribute selection, as well as transformation and cleaning of data for modeling tools.

Hypothesis and Modelling are the standard data mining steps—but in a data science project, these are not limited to statistical samples. In this phase, various modeling techniques are selected and applied, and their parameters are calibrated to optimal values. There are several techniques for the same data mining problem type. Some techniques have specific requirements on the form of data. Therefore, stepping back to the data preparation phase is often needed. A vital sub-phase is achieved here for model selection. This contains the partition of a training set for training the candidate models, and validation/test sets for evaluating model performances and selecting the best performing model, gauging model accuracy and preventing over-fitting.

Phases 2 through 4 are repeated a number of times as required. Once the understanding of data and business becomes clearer and results from initial models and hypotheses are evaluated, extra fine-tuning is achieved. These may occasionally include Step 5 and be performed in a pre-production environment before the concrete deployment.

The moment the model has been deployed in production, it is time for regular *maintenance* and *operations*. This operations phase could also follow a target model which gets well with the continuous deployment model, given the rapid time-to-market requirements in data-driven projects. The deployment contains performance tests to measure model performance, and can trigger alerts when the model performance lowers beyond a certain standard threshold.

The *optimization* phase is the final step in the data science project life cycle. This could be triggered by inadequate performance, or the need to add new data sources and retraining the model, or even to deploy a better model based on enhanced algorithms.

Real World Case 1: Retail and Advertising Insights

In order to provide effective mobile voice and data services, network operators must always acquire data on every subscriber. Besides recording the usage of mobile services for accounting and billing purposes, operators should also record each subscriber's location so it can direct calls and data streams to the cell tower to which the subscriber's handset is connected. This is how every subscriber creates a digital trail as they move around the provider network. Nowadays, the density of digital trails is amply high to correlate the combined behaviour of the subscriber crowd with characteristics of a particular location or area. For instance, it is possible to measure the allure of a place for opening a new mall, based on high-resolution analysis of how people move. Furthermore, one can also observe the impact of events, such as marketing campaigns and the launching of a competitor store, by analysing any change in traffic patterns. When gender and age group splits are entered in the data, and geo-localized data sets and social network activity are included, this segmentation adds even greater value for retailers and advertisers.

2.5 Complexity of Analytics

In general, analytics can be broken down into *three* layers: *descriptive* analytics, which can be implemented using spreadsheets or industrial strength; *predictive* analytics, which is about what will happen next; and *prescriptive* analytics, which is about how to achieve the best outcome (Akerkar 2013; Davenport and Harris 2010).

Most businesses start with descriptive analytics—the use of data to understand what happened in the past. Descriptive models quantify relationships in data in a way that is often used to classify customers or prospects into groups. Descriptive analytics prepares and analyses historical data and identifies patterns from samples for the reporting of trends. Descriptive analytics categorizes, characterizes, consolidates and classifies data. Some examples of descriptive analytics are management reports providing information about sales, customers, operations, finance and to find correlations between the various variables. For instance, Netflix uses descriptive analytics to find correlations among different movies that subscribers rent, and to improve their recommendation engine, they use historic sales and customer data. Descriptive analytics does provide significant insight into business performance and enables users to better monitor and manage their business processes. Furthermore, descriptive analytics often serves as a first step in the successful application of predictive or prescriptive analytics.

Real World Case 2: Predictive Modelling in Healthcare
Traditionally in health care, most data have been organized around the facility, the clinic or the physician—not around the patient as an individual. An organization takes steps by building a centralized data warehouse where they aggregate all data at the level of the patient. Instead of aggregating data at the hospital side, they look across different hospital wings, all physician practices and other areas where they have provided care for the patient. They create a patient's 360° view, and this gives complete knowledge about the patient. They use that data to predict risk. Some key questions to be noted here are: How many times has a patient been in emergency department in recent months? Has a patient had a colonoscopy, or has a patient had a breast test? This is required to make sure those gaps are filled. They also can look at geographical data, census data and community data to see where an individual may live and what additional risks there may be. If a patient is asthmatic, they could look and see if the patient lives in an area of higher pollution where worsening asthma may be an issue.

Furthermore, it is interesting to explore third-party consumer data. How can that data help data scientists better understand the risks of a patient, as well as actually understand more about how a patient likes to interact?

Predictive models are models of the relation between the specific performance of a unit in a sample and one or more known attributes of the unit. The purpose of the model is to assess the likelihood that a similar unit in a different sample will exhibit the specific performance. Predictive analytics uses data to find out what could happen in the future. It is a more refined and higher level usage of analytics.

Predictive analytics predicts future probabilities and trends and finds relationships in data not readily apparent with traditional analysis. Techniques such as data mining and predictive modelling reside in this space. The extensive use of data and machine learning techniques uncover explanatory and predictive models of business performance representing the inherit relationship between data inputs and outputs/outcomes. Retail and e-commerce are one of the first industries to recognize the benefits of using predictive analytics and start to employ it. In fact, understanding the customer is a first-priority goal for any retailer.

Data is at the heart of predictive analytics, and to drive a complete view, data is combined from descriptive data (attributes, characteristics, geo or demographics), behaviour data (orders, transaction, and payment history), interaction data (e-mail, chat transcripts, and Web click-streams) and attitudinal data (opinions, preferences, needs and desires). With a full view, customers can achieve higher performance such as dramatically lowering costs of claims, fighting fraud and maximizing payback, turning a call centre into a profit centre, servicing customers faster, and effectively reducing costs.

Beyond capturing the data, accessing trusted and social data inside and outside of the organization, and modelling and applying predictive algorithms, deployment of the model is just as vital in order to maximize the impact of analytics on real-time operations.

In the real world, online retailers want to understand their customer behaviour and need to maximize customer satisfaction. Retailers would like to build an optimum marketing strategy to deliver the targeted advertising, promotions and product offers to customers that will motivate them to buy. Most of the retailers find the difficulties to digest all of the data, technology, and analytics that are available to them. The ability to identify rich customers in their buying decision-making process is heavily influenced by the insights gained in the moment before the buying decision is made. Thus, to build a predictive decision making solution towards a particular product or services would be very effective to increase return on investment (ROI), conversion and customer satisfaction.

Predictive modelling should not be mistaken for data querying. The difference is that predictive modelling generally involves a machine automatically uncovering patterns and making predictions, whereas data querying involves a human interactively pulling, summarizing, and visualizing data in search of possible patterns. The former gains insights via machine, whereas the latter gains insight by a human interpreting data summaries or visuals.

Ultimately, prescriptive analytics uses data to prescribe the best course of action to increase the chances of realizing the best outcome. Prescriptive analytics evaluates and determines new ways to operate, targets business objectives and balances all constraints. Techniques such as optimization and simulation belong to this space. For example, prescriptive analytics can optimize your scheduling, production, inventory and supply chain design to deliver the right products in the right amount in the most optimized way for the right customers, on time.

Real World Case 3: Online Buyers' Behaviour Analysis
Millions of consumers visit e-commerce sites, but only a few thousand visitors may buy the products. The e-retailer wants to improve the customer experience and would like to improve the conversion rate. The purpose is to identify potential buyers based on their demographics, historical transaction pattern, clicks pattern, browsing pattern on different pages, and so on. Data analysis revealed the buyers' buying behaviours, which are highly dependent on their activities such as number of clicks, session duration, previous session, purchase session and clicks rate per session. Applying machine learning and predictive analytics methods, the tendency to conversion scores of each visitor has been derived. This leads to multiple benefits of the e-retailer offering the correct and targeted product for the customers at right time, increasing conversion rate and improving customer satisfaction. Using this analysis, the retailer can optimize the marketing strategy based on identified hidden factors of conversion, and understand the purchase funnel of customers.

2.6   From Data to Insights

Enterprise solutions are relying more and more on unstructured data for helping big companies make billion-dollar decisions. For instance, tracking the pulse of the social media data sources (such as Twitter, Facebook, Instagram, Tumbler, etc.) provides opportunities to understand individuals, groups and societies. Collecting, combining and analysing this wealth of data can help companies to harness their brand awareness, improve their product or customer service, and advertise and market their products in a better way.

Real World Case 4: Sentiment Analysis to Predict Public Response
The President Obama administration used sentiment analysis to assess public opinion on policy announcements and campaign messages ahead of the 2012 presidential election.

Now we can see how businesses are using data to treat customers more like individuals and enhance enduring relationships, by the following:

1. *Predict accurately what customers want before they ask for it*: Businesses gather a ton of data on customers, not only on what they have purchased, but also on what websites they visit, where they live, when they have communicated with customer service, and whether they interact with their brand on social media. Obviously, it is a vast amount of apparently irrelevant data, but businesses that can properly mine this can offer a more personalized touch. To properly predict the future, businesses must promote the right products to the right customers on the right channel. For example, long ago, Amazon became proficient at the recommendation of books, toys, or kitchen utensils their customers might be interested in.

2. *Get customers excited about their own data*: With the growth of wearable tools like Nike+, FuelBand, and FitBit, customers have access to more data about themselves than ever before. The food diary platform MyFitnessPal gives people not only a rundown of how many calories they have consumed each day, but also breaks down protein, fat, and carbs. However, simply providing customers data is not enough. Businesses need to filter through all of the data and extract the most relevant information as an effortless experience for customers. But if done right, data that makes a difference for customers' daily lives—whether it pertains to their health and fitness or to their money—can make a difference in a company's return on investment. Once people get hooked on their own personal data, they are more likely to continue logging in or using the product.

3. *Enhance customer service interactions*: Using data to enhance customer relationships is mainly important when customers have more channels than ever to connect with brands. For example, some airlines are using speech analytics to extract data-rich information from live-recorded interactions between customers and personnel to get a better understanding of their customers.

4. *Identify customer discomfort and help them*: Some data-driven businesses are mining into (hidden) data to solve customers' problems, and those businesses are indeed improving their customers' experience.

2.7   Building Analytics Capabilities: Case of Banking

The banking sector generates an enormous volume of data on a day-to-day basis. The traditional banking model, structured around internal product silos, is organization-centric. It prevents banks from understanding which products and services their customers have purchased across the enterprise. In this obscure environment, lending-product silos are struggling to stay profitable on a standalone basis, and the challenge will only increase.

To differentiate themselves from the competition, several banks are increasingly adopting data analytics as part of their core strategy. Analytics will be the critical game changer for these banks. Gaining insight into the dynamics of a customer's household is as important as knowing and meeting the needs of the individual. Capturing the right data exposes significant information about customers'

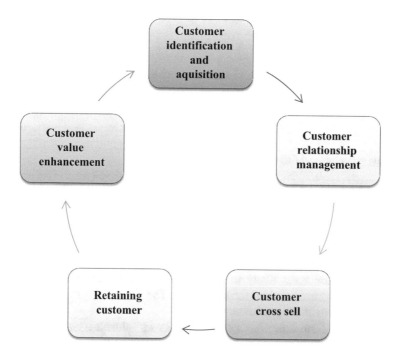

Fig. 2.4 Life cycle of banking customer

purchasing habits, financial needs, and life stages—all factors that drive their
expected purchasing decisions. As customers move through the various stages of
their lives, their financial needs change based not only on their own circumstances,
but also on those of the entire household. With banking products becoming
increasingly commoditized, analytics can help banks differentiate themselves and
gain a competitive edge.

Let us consider a typical sales process cycle in a bank, as shown in Fig. 2.4. The
sales process cycle includes knowing your target customers, determining whether
they fit into your criteria, conducting a sales call, following up and confirming
closure. Moreover, senior management would often like to monitor and use tracking
numbers. Furthermore, sales personnel know that the following information can be
very useful in various phases in the sales process:

- A list of credible customers, obtained by doing a basic check of products
  purchased and not yet purchased by them, from available data.
- A list of products most likely to be purchased by them, based on the customer's
  behavioral tendencies and buying habits.
- The likes and dislikes of the customers and their preferences.

Figure 2.4 illustrates an outlook on the typical life cycle of the banking customer
and its various phases.

Table 2.1 Information and insights provided by analytics

Phase of sales cycle	Information and insights	Activities
Customer identification and acquisition	Campaign design, acquisition analytics	Savings, current, loans
Customer relationship management	Managing portfolio, meeting transactional	Direct banking, online banking
Customer cross sell	Need analytics, family demography, credit history analysis, select more products for cross sell	Mutual funds, insurance, car loans, senior citizen accounts
Retaining customer	Churn prediction, life time value modelling	Retain customer, offer discounts on products, restructure loans
Customer value enhancement	Behavioral segmentation, product affinity modelling, differentiated pricing	Home loans, high-end structured products

Nevertheless, the use of predictive analytics to classify the products or services that customers are most likely to be interested in for their next purchase and locational intelligence provides the above data. The important question is: are there better ways to provide this information?

Analytics can furnish the senior management with valuable insights at each stage in the customer life cycle. The above table provides the type of information and insights that analytics can provide at each phase (Table 2.1).

2.8   Data Quality

Data quality is a challenge for several businesses as they look to enhance efficiency and customer interaction through data insight. Several businesses suffer from common data errors. The most common data errors are incomplete or missing data, outdated information and inaccurate data. Because of the prevalence of these common errors, the vast majority of companies suspect their contact data might be inaccurate in some way.

Data quality is a relative term, and holistically managing a data quality issues is a huge challenge for any business and is even more challenging in the sense that people have different notation about data quality while defining and validating data. There are broadly three types of data quality:

1. Data validity: Does the data do what it is supposed to do?
2. Data completeness or accuracy: Is the data good enough to do the business?
3. Data consistency: Does the data do the same thing all the time?

Once businesses are clear of data flow, what is required and expected from different systems and have clear understanding of data, they will be able to define effective test strategy. When it comes to big data, organizations are struggling to build the required infrastructure, expertise and skill set to leverage big data to positively impact their business. Unstructured data, complex business rules, complex implementation algorithms, regulatory compliance and lack of a standard big data validation approach puts an immense pressure on the independent testing team to effectively prepare for a big test effort. There is clearly a need to define data validity, data completeness and data consistency for big data validation framework to accomplish the goals and validate that data is reliable, accurate and meaningful.

In effect, data quality is the foundation for any data-driven effort. As the data proliferation continues, organizations need to prioritize data quality to ensure the success of these initiatives.

2.9 Data Preparation Process

In the next chapter, we will study various machine learning algorithms. It is important that you feed algorithms the right data for the problem you want to solve. Though you have good data, you need to make sure that it is in a useful scale and format, and that all profound features are included. The process for acquiring data ready for a machine-learning algorithm can be given in three steps:

1. Select data
   This step is concerned with selecting the subset of all available data that you will be working with. There is always a strong desire for including all available data, that the maxim 'more is better' will hold. This may or may not be true. You need to consider what data you actually need to address the question or problem you are working on. Make some assumptions about the data you require and be careful to record those assumptions so that you can test them later if needed.
2. Pre-process data
   Once you have selected the data, you need to consider how you are going to use it. This pre-processing step is about getting the selected data into a form that you can work with.

   Because of the wide variety of data sources, the collected data sets vary with respect to noise, redundancy and consistency, and it is undoubtedly a waste to store meaningless data. In addition, some analytical methods have serious requirements on data quality. Therefore, in order to enable effective data analysis, we shall pre-process data under many circumstances to integrate the data from different sources, which can not only reduce storage expense, but also improves analysis accuracy. Some relational data pre-processing techniques are considered as follows.

   Data integration: This is the basis of cutting-edge business informatics, which involves the combination of data from different sources and provides users with a uniform view of data. This is a mature research field for a traditional database. Historically, two methods have been widely recognized: data warehouse and data federation. Data warehousing includes a process called ETL (Extract,

Transform and Load). Extraction involves connecting source systems, selecting, collecting, analysing, and processing necessary data. Transformation is the execution of a series of rules to transform the extracted data into standard formats. Loading means importing extracted and transformed data into the target storage infrastructure. Loading is the most complex procedure among the three, and includes operations such as transformation, copy, clearing, standardization, screening, and data organization. A virtual database can be built to query and aggregate data from different data sources, but such a database does not contain data. On the contrary, it includes information or metadata related to actual data and its positions. Such two 'storage-reading' approaches do not satisfy the high performance requirements of data flows or search programs and applications. Compared with queries, data in such approaches is more dynamic and must be processed during data transmission. Generally, data integration methods are accompanied with flow processing engines and search engines.

Data cleaning: This is a process to identify inaccurate, incomplete, or unreasonable data, and then modify or delete such data to improve data quality. Normally, data cleaning includes five corresponding procedures such as defining and determining error types, searching and identifying errors, correcting errors, documenting error examples and error types, and modifying data entry procedures to reduce future errors.

Redundancy elimination: data redundancy refers to data repetitions or surplus, which usually occurs in many data sets. Data redundancy can increase the unnecessary data transmission expense and cause defects in storage systems, e.g., waste of storage space, leading to data inconsistency, reduction of data reliability, and data damage. Therefore, various redundancy reduction methods have been proposed, such as redundancy detection, data filtering, and data compression. Such methods may apply to different data sets or application environments. However, redundancy reduction may also bring about certain negative effects. For example, data compression and decompression cause additional computational burden. Therefore, the benefits of redundancy reduction and the cost should be carefully balanced.

3. Transform data

The final step is to transform the process data. The specific algorithm you are working with and the knowledge of the problem domain will influence this step, and you will very likely have to revisit different transformations of your pre-processed data as you work on your problem. Three common data transformations are scaling, attribute decompositions and attribute aggregations. This step is also referred to as feature engineering.

Scaling: The pre-processed data may contain attributes with mixtures of scales for various quantities such as dollars, kilograms and sales volume. Many machine-learning methods like data attributes to have the same scale, such as between 0 and 1 for the smallest and largest value for a given feature. Consider any feature scaling you may need to perform.

Decomposition: There may be features that represent a complex concept that may be more useful to a machine learning method when split into the constituent parts. An example is a date that may have day and time components, which in turn could be split out further. Perhaps only the hour of day is relevant to the problem being solved. Consider what feature decompositions you can perform.

Aggregation: There may be features that can be aggregated into a single feature that would be more meaningful to the problem you are trying to solve. For example, there may be data instances for each time a customer logged into a system that could be aggregated into a count for the number of logins, allowing the additional instances to be discarded. Consider what type of feature aggregations you could perform.

2.10 Communicating Analytics Outcomes

In data science practice, it is vital to provide decision makers with the insights they need to take action. This is possible only if managers have apt access to clear, accurate, and relevant data analytic findings in a form that aligns with their missions, needs, and prospects.

2.10.1 Strategies for Communicating Analytics

Communicating analytic outcomes may employ any of the following strategies.

- *Data Summarization* extracts crucial analytic outcomes relevant to a project's objective and aligned with the user's role and experience into a simpler and more concise form than a full report—essentially an exclusive summary. Analytic summaries should focus on the key aspects of analytic findings while enabling examination of the underlying evidence and analysis logic.
- *Synthesis of Analytic conclusions* unites seemingly separate results, whether complementary or conflicting, to explain their combined effect on the decision-making process. Given the diversity of data types and levels of accuracy, timeliness, and relevance, synthesis can be the most difficult to accomplish effectively. However, it is crucial to managers' understanding, confidence, and ability to select appropriate actions, including a decision that the data are insufficient to warrant a change to current operations.
- *Narratives* place analytic outcomes in a larger context relevant to the target audience(s), who may have roles, experience, and perspectives different from data scientists, or even leaders of the organization performing data science efforts.
- *Visualization* creates image-based static or dynamic representations of data and analytic outcomes (e.g., line charts, histograms, maps, network diagrams) to balance direct assessment of the data themselves. These visual representations may also have interactive elements that enable viewers to select data subsets,

observe data patterns over time, and otherwise perform basic analysis. Visualization requires care in selecting data types, as well as the cultural tolerance and acceptance of visuals as actionable evidence, but can be an effective method of communicating outcomes and recommendations.

2.10.2 Data Visualization

Data visualization is a representation of data that helps you see what you otherwise would have been visionless to if only looking at the naked source.

Data visualization is mostly essential for complex data analysis. In complex data, the structure, features, patterns, trends, anomalies and relationships are not easily detectable. Visualization supports the extraction of hidden patterns by presenting the data in various visual forms. Visualization not only provides a qualitative overview of large and complex data sets, it can also assist in identifying regions of interest and parameters for further quantitative analysis.

Until recently, spreadsheets have been the key tool used for data analysis and making sense of data. But when you see network data in a spreadsheet, it is hard to follow a trail—some of the information becomes invisible because it goes two or even ten levels deep. Using spreadsheets and tables to analyse complex data is too labor intensive and also relationships will remain hidden. However, when exhibited visually, data can demonstrate relationships that would have otherwise never been clear from a spreadsheet or database. In other words, by visually linking data, all relationships become more discoverable.

As an example, see the visualization in Fig. 2.5, where data from regional disaster management services were extracted and uploaded to Google maps about a storm that hit Antwerp after following a destructive path between Gent and Antwerp.

There are Semantic Graph databases that use a graph methodology for organizing and linking data. By making these graphs visual, data scientists can navigate the graph and dynamically uncover relationships when two nodes connect. Semantic graphs are optimized for both aggregated queries and pointer-chasing. With SPARQL query language and query optimizers, semantic graph databases are optimized to deal with arbitrary length strings and designed to link disparate data. The semantic graph database paradigm uses ontological systems for typing schema: large, labeled, directed graphs for data; graph pattern matching for query; and recursive query languages for graph analysis. They are mainly convenient when it comes to highly complex and large data sets.

Data scientists are uncovering new knowledge by linking recorded data with visualization through graph databases, which organize findings for comparison on a graph. With visual representations of many different data points, data scientists can then navigate the graph and dynamically uncover relationships where two nodes connect.

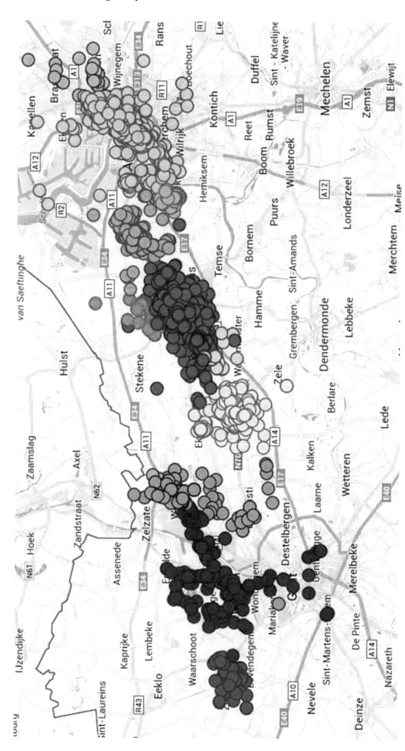

Fig. 2.5 Geographical spread data

In a business environment, visualizations can have two objectives: explanatory and exploratory. Visuals that are meant to direct you along a defined path are explanatory in nature. Various enterprise dashboards that we come across in everyday scenarios mostly fall in this category. Exploratory visuals present you many dimensions to a data set, or compare multiple data sets with each other. In this case, you can explore the visual, ask questions along the way, and find answers to those questions. Exploratory analysis can be cyclical without a precise end point. You can find many insights from a single visualization, and interact with it to gain understanding rather than make a specific decision.

2.10.3   Techniques for Visualization

The following are methods used for processing data into visualizations that uncover strategic relationships.

- *Data Regression*: can take two or more data sets and determine the level of dependency and an equation for matching. This is the mathematical equivalent of dependency modeling. Regression can determine an equation for the line and calculate how closely the data points match the line.
- *Anomaly detection*: charting techniques make deviations from the norm readily apparent. Some deviations are errors that can be removed from the data set. Others are important indicators for an important business relationship. Outlier detection identifies which points should be analysed to determine their relevance.
- *Dependency modelling*: often, two data sets will trend or cycle together because of some dependency. An evident example would be rainy weather and the sale of umbrellas. Other, less obvious relationships can be uncovered by dependency modeling. Businesses can monitor accessible factors (such as the weather) to predict less apparent factors such as sales of a certain kind of product. On charts, a positive relationship between two data sets will appear roughly as a line.
- *Clustering*: as data sets are charted, analysts can uncover a tendency for data points to cluster into groups. This can uncover data relationships in a similar fashion to dependency modeling, but for discrete variables.
- *Data Classification*: a way to use parametric data to classify entities, similar to clustering. For example, an insurance company can use 'everyday life' data for a client to determine if the client is at risk or not.

The most important takeaway is that effective data visualizations can benefit businesses as they sift through large amounts of data, in order to identify important relationships and aid in business decision making.

2.11 Exercises

1. Describe the merits and demerits of the existing analytical architecture for data scientists.
2. Consider any data-driven service you often use (for example, Amazon.com, Facebook). Envisage that you want to store the data for that service in a simple spreadsheet. What would each row in the spreadsheet represent? What would some of the data columns be?
3. Explore the following data visualization types:

   - Tag clouds
   - Clustergram
   - Motion charts
   - Dashboard

4. (*Project*) Analyse real world data science problems, identify which methods are appropriate, organize the data appropriately, apply one or more methods, and evaluate the quality of the solution.
5. (*Project*) Consider the following business cases and describe how the business should respond to the issues and problems identified in each case. Conclude your analysis by reviewing your findings.

   (a) The dairy farmers wants to know if they are getting reasonable market value pricing for their milk and if structural factors within the regions and the cooperatives could be changed to improve the farmers' economic standing going forward. Assume that the price that farmers receive for their raw milk is regulated. This guarantees that farmers receive a minimum price for the milk sold to processing plants. In some regions, however, the minimum is not a binding price floor, and the market price is above the minimum. The difference between the minimum price and the market price is commonly referred to as the over-order premium. The dairy farmers are eligible for an over-order premium depends on factors including:

      - Number and production capabilities of dairy cows physically located in the region
      - Dairy processing capacity and capacity utilization
      - Ability to profitably import raw milk from outside the region
      - Minimum price formulae
      - Variations in the market price for finished dairy products
      - Weather conditions

   (b) The supermarket wants to optimize its shelf space allocation for an entire product category for better profitability. In particular, the oil and shortening product category is not obtaining adequate results for the supermarket chain as compared to the shelf space provided. Transactional data for the entire category over a period of several years is available and there is a demand to maximize sales while minimizing inventory investment.

References

Akerkar, R. (2013). *Big data computing.* Boca Raton.: Chapman and Hall/CRC.

Davenport, T., & Harris, J. (2010). *Analytics at work: Smarter decision, better results.* Boston: Harvard Business Review Press.

Manna, M, (2014). *The business intelligence blog* [Internett]. Available at: https://biguru.wordpress.com/2014/12/22/the-data-science-project-lifecycle/. Funnet January 2016.

Shearer, C. (2000). The CRISP-DM model: The new blueprint for data mining. *Journal of Data Warehousing, 5,* 13–22.

Chapter 3
Basic Learning Algorithms

3.1 Learning from Data

This chapter provides a broad yet methodical introduction to the techniques and practice of machine learning. Machine learning can be used as a tool to create value and insight to help organizations to reach new goals. We have seen the term 'data-driven' in earlier chapters and have also realised that the data is rather useless until we transform it into information. This transformation of data into information is the *rationale* for using machine learning.

Learning is assimilated with terms such as *to gain knowledge or understanding, by study, instruction, or experience*. To understand how humans learn, researchers attempt to develop methods for accomplishing the acquisition and application of knowledge algorithmically, termed as *machine learning*. In simple words, machine learning is an algorithm that can learn from data without relying on rules-based programming. Machine learning can be viewed as general inductive process that automatically builds a model by learning the inherent structure of a data set depending on the characteristics of the data instances. Over the past decade, machine learning has evolved from a field of laboratory demonstrations to a field of significant commercial value.

Developing computer programs that 'learn' requires knowledge from many fields. The discipline of machine learning integrates many distinct approaches, such as probability theory, logic, combinatorial optimization, search, statistics, reinforcement learning and control theory. The developed methods are at the basis of many applications, ranging from vision to language processing, forecasting, pattern recognition, games, data mining, expert systems and robotics.

But the basic question is: why would machines learn, when they can be designed from the beginning to perform as desired? Apart from the reason that it may provide explanations about how humans learn, there are important engineering reasons for machines to learn. Nilsson (1996) mentions the following lessons:

© Springer International Publishing Switzerland 2016
R. Akerkar, P.S. Sajja, *Intelligent Techniques for Data Science*,
DOI 10.1007/978-3-319-29206-9_3

- Some tasks cannot be defined well except by example; that is, we might be able to specify input/output pairs, but not concise relationships between inputs and the desired outputs. We would like machines to be able to adjust their internal structure to produce correct outputs for a large number of sample inputs, and thus suitably constrain their input/output function to approximate the relationship implicit in the examples.
- It is possible that important relationships and correlations are hidden in large piles of data. Machine learning methods can often be used to extract these relationships.
- Human designers often produce machines that do not work as well as desired in the environments in which they are used. In fact, certain characteristics of the working environment might not be completely known at design time. Machine learning methods can be used for on-the-job improvement of existing machine designs.

Real World Case 1: Improving Healthcare Delivery
For hospitals, patient readmission is a critical issue, and not just out of concern for the patient's health and welfare. Public and private insurers penalize hospitals with a high readmission rate, so hospitals have a financial stake in making sure they discharge only those patients who are well enough to stay healthy. Thus, prominent healthcare systems use machine learning to construct risk scores for patients, which case administrators work into their discharge decisions. This supports better utilization of nurses and case administrators, prioritizing patients according to risk and complexity of the case.

The following terms are often used in the field of machine learning:

- *Instance*: An instance is an example in the training data. An instance is described by a number of attributes. One attribute can be a class label.
- *Attribute*: An attribute is an aspect of an instance (e.g., temperature, humidity). Attributes are often called features in machine learning. A special attribute is the class label that defines the class this instance belongs to.
- *Classification*: A classifier is able to categorize given instances (test data) with respect to pre-determined or learned classification rules.
- *Training*: A classifier learns the classification rules based on a given set of instances (training data). Some algorithms have no training stage, but do all the work when classifying an instance.
- *Clustering*: The process of dividing an unlabelled data set (no class information given) into clusters that encompass similar instances.

Data scientists use many different machine learning tools to perform their work, including analytic languages such as R, Python, and Scala.

Efficient machine learning procedures include the following features:

- Ability to treat missing data
- Ability to transform categorical data
- Regularization techniques to manage complexity
- Grid search capability for automated test and learn
- Automatic cross-validation

One crucial question is how to select an appropriate machine learning algorithm in the context of certain requirements. The answer indeed depends on the size, quality, and the nature of the data. It depends what you want to do with the solution.

- *Accuracy*: Whether obtaining the best score is the aim or an approximate solution while trading off overfitting. You may be able to cut your processing time considerably by sticking with more approximate methods.
- *Training time*: The amount of time available to train the model. Some algorithms are more sensitive to the number of data points than others. When time is limited and the data set is large, it can drive the choice of algorithm.
- *Linearity*: Linear classification algorithms assume that classes can be separated by a straight line. Though these assumptions are good for certain problems, they bring accuracy down on some.
- *Number of parameters*: Parameters, such as error tolerance or number of iterations, or options between variants of how the algorithm performs, affect an algorithm's behaviour. The training time and accuracy of the algorithm can occasionally be rather sensitive to getting just the right settings.
- *Number of features*: For certain types of data, the number of features can be very large compared to the number of data points. The large number of features can overload some learning algorithms, making training speed incredibly slow.

Moreover, some learning algorithms make particular assumptions about the structure of the data or the desired results.

Machine learning can be broken down into a few categories. The three most popular are supervised, unsupervised and reinforcement learning. We will discuss these categories in the next sections. We refer to Akerkar and Lingras (2007), Witten and Frank (2005) and Mitchell (1997) for a complete discussion of various learning algorithms. The R (programming) code for some of these methods is provided in Chap. 9.

3.2   Supervised Learning

In supervised learning, the model defines the effect one set of observations (inputs) has on another set of observations (outputs). In other words, the inputs are assumed to be at the beginning and outputs at the end of the causal chain. The models can include mediating variables between the inputs and outputs.

In short, supervised learning algorithms make predictions based on a set of examples. For example, if we work with historical data from a marketing campaign, we can classify each impression by whether or not the prospect responded, or we can determine how much they spent. Supervised techniques provide powerful tools for *prediction* and *classification*.

3.2.1 Linear Regression

Algorithms that develop a model based on equations or mathematical operations on the values taken by the input attributes to produce a continuous value to represent the output are called of regression algorithms. We usually use regression techniques when we want to provide ourselves with the option for optimizing our results. The input to these algorithms can take both continuous and discrete values depending on the algorithm, whereas the output is a continuous value.

Let us understand linear regression: Assume that you ask a student in primary school to arrange all students in her class by increasing order of weight, without asking them their weights. What do you think the student will do? She would visually analyse the height and physique of students and arrange them using a combination of these visible parameters. Here, the student has actually figured out that height and physique are correlated to weight by a relationship, which looks like the following equation.

Linear regression is used to estimate real values based on continuous variable(s). Here, we establish relationship between independent and dependent variables by fitting a best line. This best fit line is known as a regression line and is represented by the linear equation

$$Y = a * X + b$$

where, Y is dependent variable, a is slope, X is independent variable, and b is an intercept.

The coefficients a and b are derived based on minimizing the sum of squared difference of distance between data points and regression line.

Real World Case 2: Forecasting
In hotel business, customer transactional data can be useful in the development of a forecasting model that accurately produces meaningful expectations. Irrespective of whether a hotel chain relies on moving average or time series forecasting algorithms, machine learning can improve the statistical reliability of forecast modelling. Estimating in advance how much of and when menu items need to be prepared is critical to efficient food production

(continued)

management. Regression models can provide a prediction of product usage by part of day, given available sales data. Also, knowing how much product was sold during any period can be useful in supporting an efficient inventory replenishment system that minimizes the amount of capital tied up in stored products.

The linear regression algorithm gives the best results when there is some linear dependency among the data. It requires the input attributes and target class to be numeric and does not allow missing attributes values. The algorithm calculates a regression equation to predict the output (x) for a set of input attributes $a_1, a_2, \ldots, a_k$. The equation to calculate the output is expressed in the form of a linear combination of input attributes with each attribute associated with its respective weight $w_0, w_1, \ldots, w_k$, where w_1 is the weight of a_1 and a_0 is always taken as the constant 1. An equation takes the form

$$x = w_0 + w_1 a_1 + \ldots\ldots\ldots + w_k a_k.$$

The weights must be selected in such a manner that they minimize error. To obtain better accuracy, higher weights must be assigned to those attributes that influence the result the most. A set of training instances is used to update the weights. At the start, the weights can be assigned random values or all be set to a constant (such as 0). For the first instance in the training data, the predicted output is obtained as

$$w_0 + w_1 a_1^{(1)} + \ldots\ldots\ldots + w_k a_k^{(1)} = \sum_{j=0}^{k} w_j a_j^{(1)},$$

where the superscript for attributes gives the instance position in the training data. After the predicted outputs for all instances are obtained, the weights are reassigned so as to minimize the sum of squared differences between the actual and predicted outcome. Thus the aim of the weight update process is to minimize

$$\sum_{i=1}^{n} \left(x^{(i)} - \sum_{j=0}^{k} w_j a_j^{(i)} \right),$$

which is the sum of the squared differences between the observed output for the ith training instance $(x^{(i)})$ and the predicted outcome for that training instance obtained from the linear regression equation.

In the given Fig. 3.1, we have identified the best-fit line having the linear equation $y = 0.2741x + 12.8$ Using this equation, we can find the weight, knowing the height of a person.

Fig. 3.1 Linear regression

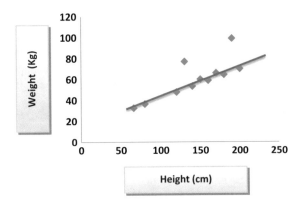

There are two kinds of linear regression.

- Simple linear regression
- Multiple linear regression

Consider an analyst who wishes to establish linear relationship between the daily change in a company's stock prices and other explanatory variables, such as the daily change in trading volume (number of shares traded in a security or an entire market during a given period of time) and the daily change in market returns. If she runs a regression with the daily change in the company's stock prices as a dependent variable and the daily change in trading volume as an independent variable, this would be an example of a simple linear regression with one explanatory variable. If the analyst adds the daily change in market returns into the regression, it would be a multiple linear regression.

If we are using regression with big data, we are presumably looking for correlations and some data providence.

3.2.2   Decision Tree

Decision tree is a supervised learning method. Decision trees are powerful and popular tools for classification and prediction. The beauty of tree-based methods is due to the fact that decision trees represent *rules*. Rules can readily be expressed so that humans can understand them. They can also be expressed in a database language like SQL, so that records falling into a particular category may be retrieved.

The algorithm works for both categorical and continuous dependent variables. In this algorithm, you split the population into two or more homogeneous sets. This is done based on most significant attributes/independent variables to make as distinct groups as possible.

In some applications, the accuracy of a classification or prediction is the only thing that matters. In certain circumstances, the ability to explain the reason for a decision is crucial. In health insurance underwriting, for example, there are legal prohibitions against discrimination based on certain variables. An insurance company could find itself in the position of having to demonstrate to the satisfaction of a court of law that it has not used illegal discriminatory practices in granting or denying coverage. There are a variety of algorithms for building decision trees that share the desirable trait of explicability.

Let us consider Fig. 3.2, in which you can see that population is classified into four different groups based on multiple attributes to identify 'if they will play or not'.

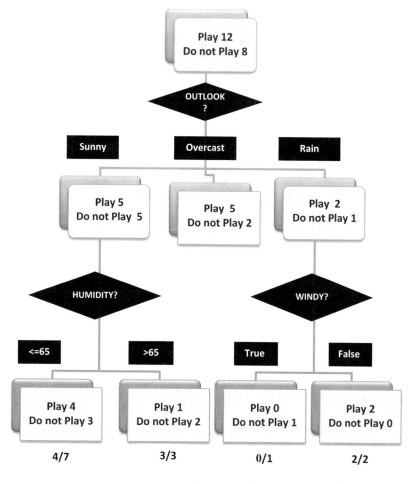

Note: Figures below the leaf nodes indicate scores

Fig. 3.2 Decision tree

The decision tree construction process is concerned with identifying the splitting attributes and splitting criteria at every level of the tree. The goal of the decision tree construction process is to generate simple, logical rules with high accuracy. Occasionally, the classification efficiency of the tree can be enhanced by altering the tree through pruning and grafting. These processes are activated after the decision tree is constructed. The following are some of the desired properties of decision-tree–generating methods:

- The methods should be able to handle both numerical and the categorical attributes, and
- the methods should provide a clear indication of which fields are most important for prediction or classification.

Demerits of the decision trees are:

- Certain decision trees can only deal with binary-valued target classes, and the procedure of developing a decision tree is computationally expensive. At each node, each candidate splitting field is examined before the best split can be found.

Real World Case 3: Reducing Churn and Risk
An insurance company has recognized that they had higher than normal mid-term cancellations of home insurance policies. Reducing customer churn and risk is a key variable for the company's success. So, the data scientist has created, tested and refined a model that predicted the cancellation of its mid-term policy, looking at 13 days after inception and 27 days before renewal on a segmented data set.

All the decision tree construction techniques are based on the principle of recursively partitioning the data set until homogeneity is achieved. The construction of the decision tree involves the following three major phases.

- *Construction phase.* The initial decision tree is constructed in this phase, on the entire training data set. It requires recursively partitioning the training set into two or more subsets using a splitting criterion, until a stopping criterion is met.
- *Pruning phase.* The tree constructed in the previous phase may not result in the best possible set of rules due to overfitting. The pruning phase removes some of the lower branches and nodes to improve its performance.
- *Processing phase.* The pruned tree is further processed to improve understand-ability.

Though these three phases are common to most well-known algorithms, some algorithms attempt to integrate the first two phases into a single process.

3.2.2.1 M5Prime Algorithm

The M5P or M5Prime algorithm is a reconstruction of Quinlan's M5 algorithm for inducing trees of regression models (Quinlan 1986). M5P combines a conventional decision tree with the possibility of linear regression functions at the nodes.

First, a decision-tree induction approach is used to build a tree, but instead of maximizing the information gain at each inner node, a splitting criterion is used to minimize the intra-subset variation in the class values down each branch. The splitting procedure in M5P stops if the class values of all instances that reach a node vary very slightly, or if only a few instances remain.

Second, the tree is pruned back from each leaf. When pruning, an inner node is turned into a leaf with a regression plane.

Third, to avoid sharp discontinuities between the subtrees, a smoothing procedure is applied that combines the leaf model prediction with each node along the path back to the root, smoothing it at each of these nodes by combining it with the value predicted by the linear model for that node.

M5 constructs a tree to predict values for a given instance. The algorithm needs the output attribute to be numeric while the input attributes can be either discrete or continuous. At each node in the tree, a decision is made to pursue a particular branch based on a test condition on the attribute associated with that node. Each leaf has a linear regression model associated with it of the form

$$w_o + w_1 a_1 + \ldots\ldots\ldots + w_k a_k,$$

based on some of the input attributes $a_1, a_2, \ldots, a_k$ in the instance and whose respective weights $w_0, w_1, \ldots, w_k$ are calculated using standard regression. When the leaf nodes contain a linear regression model to obtain the predicted output, the tree is called a model tree.

To construct a model tree (see Fig. 3.3), using the M5 algorithm, we start with a set of training instances. The tree is created using a divide-and-conquer technique.

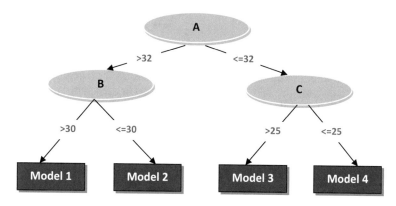

Fig. 3.3 M5 model tree for predicting temperature

At a node, starting with the root node, the instance set that reaches it is either associated with a leaf or a test condition is chosen that splits the instances into subsets based on the test outcome. A test is based on an attributes value, which is used to decide which branch to follow. There are several potential tests that can be used at a node. In M5, the test that maximizes error reduction is used. For a test, the expected error reduction is found using

$$\Delta \text{error} = \text{stdev}(S) - \sum_i \left(\frac{|S_i|}{|S|} \text{stdev}(S_i) \right),$$

where S is the set of instance passed to the node, *stdev(S)* is its standard deviation, and S_i is the subset of S resulting from splitting at the node with the *ith* outcome for the test. This process of creating new nodes is repeated until a there are too few instances to proceed further, or the variation in the output values in the instances that reach the node is small.

As soon as the tree has been formed, a linear model (regression equation) is created at each node. The attributes used in the equation are those that are tested or are used in linear models in the subtrees below the node. The attributes tested above this node are not used in the equation, as their effect on predicting the output has already been acquired in the tests done at the above nodes. The linear model that is built is further simplified by eliminating attributes in it. The attributes whose removal from the linear model leads to a reduction in error are eliminated. Error is defined as the complete difference between the output value predicted by the model and the real output value seen for a given instance.

The tree constructed can take a complex form. Pruning plays a significant role in model fitting of any data set. The tree is pruned so as to make it simpler without losing the basic functionality. Starting from the bottom of the tree, the error is calculated for the linear model at each node. If the error for the linear model at a node is less than the model subtree below, then the subtree for this node is pruned. In the case of missing values in training instances, M5P changes the expected error reduction equation to

$$\Delta \text{error} = \frac{m}{|S|} * \beta(i) * \left[\text{stdev}(S) - \sum_i \left(\frac{|S_i|}{|S|} \text{stdev}(S_i) \right) \right],$$

where m is the number of instances without missing values for that attribute, S is the set of instances at the node, $\beta(i)$ is the factor multiplied in case of discrete attributes, and j takes values L and R, with S_L and S_R being the sets obtained from splitting at that attribute.

During testing, an unknown attribute value is replaced by the average value of that attribute for all training instances that reach the node, with the effect of always choosing the most populated subnode.

3.2.2.2 ID3 Decision Tree Algorithm

ID3 (Quinlan 1986) represents concept as *decision trees*. A decision tree is a classifier in the form of a tree structure where each node is either:

- a *leaf node*, indicating a class of instances, or
- a *decision node*, which specifies a test to be carried out on a single attribute value, with one branch and a subtree for each possible outcome of the test.

A decision tree can be used to classify an instance by starting at the root of the tree and moving through it down to a leaf node, which provides the classification of the instance.

A decision tree classifies a given instance by passing it through the tree starting at the top and moving down until a leaf node is reached. A typical decision tree predicting the current temperature is shown in Fig. 3.4. The value at that leaf node gives the predicted output for the instance. At each node, an attribute is tested and the branches from the node correspond to the values that attribute can take. When the instance reaches a node, the branch taken depends on the value it has for the attribute being tested at the node.

The ID3 algorithm constructs a decision tree based on the set of training instances given to it. It takes a greedy top-down approach for the construction of the tree, starting with the creation of the root node. At each node, the attribute that best classifies all the training instances that have reached that node is selected as the test attribute. At a node, only those attributes not used for classification at other nodes above it in the tree are considered. To select the best attribute at a node, the

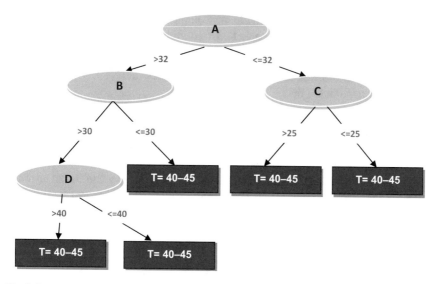

Fig. 3.4 A decision tree to predict the current temperature at location C based on temperature readings taken from a set of nearby places

information gain for each attribute is calculated and the attribute with the highest information gain is selected. Information gain for an attribute is defined as the reduction in *entropy* caused by splitting the instances based on values taken by the attribute. The information gain for an attribute A at a node is calculated using

$$\text{InformationGain}\,(S, A) = \text{Entropy}(S) - \sum_{v \in \text{Values}(A)} \left(\frac{|S_v|}{|S|} \text{Entropy}(S) \right),$$

where S is the set of instances at that node and $|S|$ is its cardinality, S_v is the subset of S for which attribute A has value v, and entropy of the set S is calculated as

$$\text{Entropy}(S) = \sum_{i=1}^{\text{numclasses}} -p_i \log_2 p_i,$$

where p_i is the proportion of instances in S that have the *ith* class value as output attribute.

A new branch is inserted below the node for each value taken by the test attribute. The training instances that have the test attribute value associated with the branch taken are passed down the branch, and this subset of training instances is used for the creation of further nodes. If this subset of training instances has the same output class value then a leaf is generated at the branch end, and the output attribute is assigned that class value. In the case where no instances are passed down a branch then a leaf node is added at the branch end that assigns the most common class value in the training instances to the output attribute. This process of generating nodes is continued until all the instances are properly classified or all the attributes have been used or when it's not possible to divide the examples.

Extensions were added to the basic ID3 algorithm to

1. deal with continuous valued attributes,
2. deal with instances that have missing attribute values, and
3. prevent overfitting the data.

When a discrete valued attribute is selected at a node the number of branches formed is equal to the number of possible values taken by the attribute. In the case of a continuous valued attribute two branches are formed based on a threshold value that best splits the instances into two.

There may arise cases where an instance has no (missing) value for an attribute or has an unknown attribute value. The missing value can be replaced by the most common value for that attribute among the training instances that reach the node where this attribute is tested.

C4.5 is an algorithm used to generate a decision tree developed by Quinlan (Quinlan 1993). It is an extension of ID3 algorithm. In C4.5 algorithm, the probability for each possible value taken by the attribute with missing value is calculated, based on the number of times it is seen in the training instances at a node. The probability values are then used for calculation of information gain at the node.

In the ID3 algorithm, sometimes due to too small of a training set being used, the tree built correctly classifies the training instances but fails when applied on the entire distribution of data because it focuses on the spurious correlation in the data when the remaining amount of data is small; this is known as *overfitting*. To avoid overfitting, C4.5 uses a technique called rule-post pruning. In rule post-pruning, the tree is converted into a set of rules after it is constructed.

From each rule generated for the tree, those antecedents are pruned which do not reduce the accuracy of the model. Accuracy is measured based on the instances present in the validation set, which is a subset of the training set not used for building the model.

Real World Case 4: Finding New Customers
Finding new customers is a common need for any business. When a prospective customer visits company's website, there are many products (different insurance plans) from which to choose. With the web and with the advantage of targeted marketing by means of machine learning models, many data-driven businesses have experienced rise in new customer acquisition via online interactions. This is profitable especially because the costs involved online are less than direct contact using regular posts or mails.

3.2.3 Random Forest

Random Forest is a hallmark phrase for an ensemble of decision trees. In Random Forest, we have collection of decision trees, so known as 'Forest'.

Decision tree is encountered with overfitting problem and ignorance of a variable in case of small sample size and large p value; whereas, random forests are well suited to small sample size and large p value problems. Random forest comes at the expense of some loss of interpretability, but boosts the performance of the ultimate model.

There are two well-known methods of classification trees, namely, boosting and bagging. In boosting, successive trees give extra weight to points incorrectly predicted by earlier predictors. In the end, a weighted vote is taken for prediction. In bagging, successive trees do not depend on earlier trees—each is independently constructed using a bootstrap sample of the data set. Eventually, a simple majority vote is taken for prediction.

(Breiman 2001) proposed random forests, which add an additional layer of randomness to bagging. In addition to constructing each tree using a different bootstrap sample of the data, random forests change how the classification or regression trees are constructed. In standard trees, each node is split using the best split among all variables. In a random forest, each node is split using the best among a subset of predictors randomly chosen at that node (Breiman et al. 1984).

The random forests algorithm is as follows:

1. Draw n_{tree} bootstrap samples from the original data.
2. For each of the bootstrap samples, grow an unpruned classification or regression tree, with the following modification: at each node, rather than choosing the best split among all predictors, randomly sample m_{try} of the predictors and choose the best split from among those variables. (Bagging can be thought of as the special case of random forests obtained when $m_{try} = p$, the number of predictors.)
3. Predict new data by aggregating the predictions of the n_{tree} trees (i.e., majority votes for classification, average for regression).

An estimate of the error rate can be obtained, based on the training data, by the following:

1. At each (bootstrap) iteration, predict the data not in the bootstrap sample using the tree grown with the bootstrap sample.
2. Aggregate the out-of-bag (out-of-bag is equivalent to validation or test data. In random forests, there is no need for a separate test set to validate result) predictions and calculate the error rate, and call it the out-of-bag estimate of error rate.

However, the obvious question to ask is why does the ensemble work better when we choose features from random subsets rather than learn the tree using the traditional algorithm? The ensembles are more effective when the individual models that comprise them are uncorrelated. In traditional bagging with decision trees, the constituent decision trees may end up being very correlated, because the same features will tend to be used repeatedly to split the bootstrap samples. By limiting each split-test to a small, random sample of features, we can decrease the correlation between trees in the ensemble. Besides, by confining the features we consider at each node, we can learn each tree much faster, and hence, can learn more decision trees in a given amount of time. Therefore, not only can we build many more trees using the randomized tree learning algorithm, but these trees will also be less correlated. For these reasons, random forests tend to have excellent performance.

3.2.4 k-Nearest Neighbour

We use nearest neighbour algorithm when we have many objects to classify and the classification process is tedious and time consuming. This method is suitable when it makes good business cognizance to do so.

The k-nearest neighbour algorithm (kNN) is a method for classifying objects based on closest training examples in the feature space. It is a type of instance-based or lazy learning where the function is only approximated locally and all computation is delayed until classification. K-nearest neighbour is a simple algorithm that stores all available cases and classifies new cases by a majority vote of its k neighbours.

The case being assigned to the class is most common amongst its k nearest neighbours measured by a distance function.

> **Real World Case 3: Credit Rating in Banking Sector**
> A bank has a database of customers' details and their credit rating. These details would probably be the individual's financial characteristics, such as how much they earn, whether they own or rent a house, and so on, and would be used to calculate the individual's credit rating. However, the process for calculating the credit rating from her/his details is relatively expensive, so the bank would like to find some way to reduce this cost. They recognize that by the very nature of a credit rating, customers who have similar financial details would be given similar credit ratings. As a result, they would like to be able to use this existing database to predict a new customer's credit rating, without having to perform all the calculations.

Let us construct a classification method using no assumptions about the form of the function, $y = f(x_1, x_2, \ldots x_p)$, that relates the dependent (or response) variable, y, to the independent (or predictor) variables $x_1, x_2, \ldots x_p$. The only assumption we make is that it is a 'smooth' function. This is a non-parametric method, because it does not involve estimation of parameters in an assumed function form such as the linear form we encountered in linear regression.

We have training data in which each observation has a y value, which is just the class to which the observation belongs. For example, if we have two classes, y is a binary variable. The idea in k-Nearest Neighbour methods is to dynamically identify k observations in the training data set that are similar to a new observation, say

$$(u_1, u_2, \ldots up),$$

that we wish to classify and to use these observations to classify the observation into a class, $\hat{u}$. If we knew the function f, we would simply compute

$$\hat{u} = f(u_1, u_2, \ldots u_p).$$

If all we are prepared to assume is that f is a smooth function, a reasonable idea is to look for observations in our training data that are near it and then to compute $\hat{u}$ from the values of y for these observations.

Now the Euclidean distance between the points $(x_1, x_2, \ldots x_p)$ and $(u_1, u_2, \ldots u_p)$ is

$$\sqrt{(x_1 - u_1)^2 + (x_2 - u_2)^2 + \cdots + (x_p - u_p)^2}$$

We will examine other ways to define distance between points in the space of predictor variables when we discuss clustering methods.

The simplest case is $k = 1$ where we find the observation that is the nearest neighbour and set $\hat{u} = y$ where y is the class of the nearest neighbour. If we have a large amount of data and used an arbitrarily classification rule, at best we would be able to reduce the misclassification error to half that of the simple 1-NN rule.

For k-NN, we extend the idea of 1-NN as follows. Find the nearest k neighbours and then use a majority decision rule to classify a new observation. The advantage is that higher values of k provide smoothing, which reduces the risk of overfitting due to noise in the training data. In typical applications, k is in units or tens. Notice that if $k = n$, the number of observations in the training data set, we are merely predicting the class that has the majority in the training data for all observations, irrespective of the values of $(u_1, u_2, \ldots u_p)$. This is clearly a case of over smoothing unless there is no information at all in the independent variables about the dependent variable.

The k-nearest neighbour algorithm is sensitive to the local structure of the data. The best choice of k depends on the data; generally, larger values of k reduce the effect of noise on the classification, but make boundaries between classes less distinct. A good k can be selected by various heuristic techniques such as cross-validation. The accuracy of the algorithm can be severely degraded by the presence of noisy or irrelevant features, or if the feature scales are not consistent with their importance. An example of kNN is shown in Fig. 3.5.

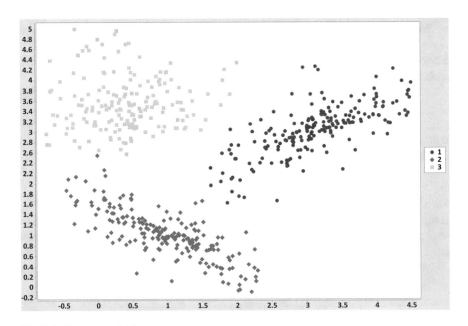

Fig. 3.5 *K*-nearest neighbour

3.2.5 Logistic Regression

Logistic regression is basically a classification not a regression algorithm. Suppose we are interested in the circumstances that influence whether a political candidate wins an election or not. The outcome (response) variable is binary (0/1): win or lose. The predictor variables of interest are the amount of money spent on the campaign, the amount of time spent campaigning negatively and whether or not the candidate is an incumbent. These are the explanatory or independent variable. Here logistic regression applies in a situation where we want to find out the probability associated with any two outcomes.

Logistic regression is used to estimate discrete values based on a given set of independent variable(s). It predicts the probability of occurrence of an event by fitting data to a logit function. Hence, it is also known as logistic regression. Since it predicts the probability, its output values lies between 0 and 1.

Real World Case 4: Predicting Business Bankruptcy
Creditors and shareholders in businesses need to be able to predict the probability of default for profitable business decisions. In the financial sector, precise assessment of the probability of bankruptcy can lead to better lending practices and superior fair value estimates of interest rates that reflect credit risks. For instance, accounting firms may risk grievances if the auditors fail to issue an early warning. Traditionally, the credit or counterparty risk assessment was to merely use ratings issued by the standard credit rating agencies. As many investors have discovered in recent times, these ratings tend to be reactive rather than predictive. Hence, there is an immense demand to develop accurate quantitative models for prediction of business bankruptcy. A key methodology to develop quantitative models for such prediction has been to learn the relationship of default with firm variables from data using statistical models. In such practice, models based on logistic regression, multivariate discriminant analysis, and neural networks have been used to predict business bankruptcy.

To obtain a better understanding of what the logit function is, we will now introduce the notation of odds. The odds of an event that occurs with probability p is defined as:

$$\text{odds} = \frac{p}{(1-p)} = \frac{\text{probability of event occurrence}}{\text{probability of not event occurrence}}$$

$$ln\,(\text{odds}) = ln\,(p/\,(1-p))$$

$$\text{logit}(p) = ln\,(p/\,(1-p)) = b_0 + b_1 X_1 + b_2 X_2 + b_3 X_3 .... + b_k X_k$$

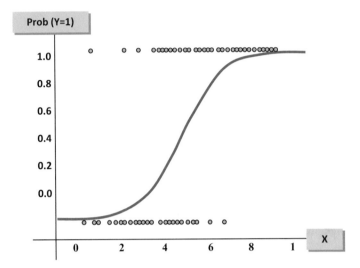

Fig. 3.6 Logistic regression (the logit function)

Above, *p* is the probability of presence of the characteristic of interest. It chooses parameters that maximize the likelihood of observing the sample values rather than those that minimize the sum of squared errors, just as in ordinary regression. For the sake of simplicity, let us simply say that this is one of the best mathematical ways to replicate a step function.

Figure 3.6 shows how the `odds` function looks. As we can see, the odds for an event are not bounded and go from 0 to infinity when the probability for that event goes from 0 to 1.

However, it is not always very intuitive to think about `odds`. Even worse, `odds` are quite unappealing to work with due to asymmetry.

Logistic regression has several advantages over linear regression, especially that it is more robust and does not assume a linear relationship since it may handle non-linear effects. However it requires much more data to achieve stable, meaningful results. Some other advantages are:

• Good accuracy for many simple data sets.
• Makes no assumptions about distributions of classes.
• Resistant to overfitting.
• Easily extended to multiple classes.

3.2.6 Model Combiners

The term *model combiners* (ensemble methods) is basically retained for bundled fits produced by a stochastic algorithm, the output of which is some combination of a large number of passes through the data. Such methods are loosely related to iterative procedures, on one hand, and to bootstrap procedures, on the other. Various commentators have called this approach most significant development in machine learning over the last decade. Ensemble methods had been widely applied in many fields such as web ranking algorithm, classification and clustering, time series and regression problems, and water quality application, among others.

Building predictive models has traditionally involved the training of many models from the entire training data set, and via various validation techniques, picking the best performing candidate. In reality, however, a single model may well be unstable and does not capture all the useful patterns that might be found in the data. Model combiners exploit some of the weaknesses of a single model approach, and model instability can be used to good effect. There are other issues, such as the inclusion of random modifications to the learning algorithm, which result in models that are unusually accurate.

A number of techniques have evolved to support ensemble learning, the best known being bagging (bootstrap aggregating) and boosting (Oza 2004). A number of differences exist between bagging and boosting. Bagging uses a majority voting scheme, while boosting uses weighted majority voting during decision-making. Classifiers generated by boosting are independent of each other, while those of bagging are dependent on each other. In bagging, instances selected to train base classifiers were bootstrapped duplicate copies of the training data, which implies that each instance has the same chance of being in each training data set. In boosting, the training data set of each subsequent classifier gradually concentrates on instances misclassified by earlier trained classifiers. But bagging has the advantage over boosting in that it reduces variance and minimizes error.

3.2.6.1   AdaBoost Algorithm

Boosting is a forward stage-wise additive model. A boosted approach is taken where we have enormous data to make a prediction and we pursue exceptionally high predictive power. The boosting combines multiple weak predictors to a build strong predictor.

AdaBoost is a meta-algorithm in the sense that it is used to increase the performance of other algorithms. It assigns a *weight* to each training example, which determines the probability that each example should appear in the training set. Examples with higher weights are more likely to be included in the training set, and vice versa. After training a classifier, AdaBoost increases the weight on the misclassified examples so that these examples will make up a larger part of the next classifiers training set, and the next classifier trained will perform better on them.

Let us understand the concept of AdaBoost that generates a sequence of base models with different weight distributions over the training set. Its inputs are a set of N training examples, a base model learning algorithm L_b, and the number M of base models that we wish to combine. AdaBoost was originally designed for two-class classification problems; therefore, for this explanation we will assume that there are two possible classes. However, AdaBoost is regularly used with a larger number of classes. The first step in AdaBoost is to construct an initial distribution of weights D_1 over the training set. This distribution assigns equal weight to all N training examples.

The AdaBoost algorithm is illustrated as follows:

$AdaBoost\ (\{(x_1, y_1), (x_2, y_2), \ldots, (x_N, y_N)\}, L_b, M)$

Initialize $D_1(n) = \frac{1}{N}$ for all $n \in \{1, 2, \ldots, N\}$

For each $m = 1, 2, \ldots, M$:

$h_m = L_b\ (\{(x_1, y_1), (x_2, y_2), \ldots, (x_N, y_N)\}, D_m)$

$\varepsilon_m = \sum_{n:h_m(x_n)y_n} D_m(n)$

If $\varepsilon_m \geq 1/2$ then,

set $M = m - 1$ and abort this loop

Update distribution D_m :

$$D_{m+1}(n) = D_m(n) \times \begin{cases} \frac{1}{2(1-\varepsilon_m)} & \text{if } h_m(x_m)=y_m \\ \frac{1}{2\varepsilon_m} & \text{Otherwise} \end{cases}$$

$$\text{Return } h_{fin}(x) = \arg\max_{y \in Y} \sum_{m-1}^{M} I\ (h_m(x) = y) \log\left(\frac{1 - \varepsilon_m}{\varepsilon_m}\right)$$

To construct the first base model, we call L with distribution D over the training set.

After getting back a model h_1, we calculate its error ε_1 on the training set itself, which is just the sum of the weights of the training examples that h_1 misclassifies.

We require that $\varepsilon_1 < \frac{1}{2}$. This is the *weak learning* assumption; the error should be less than what we would achieve through randomly guessing the class. If this condition is not satisfied, then we stop and return the ensemble consisting of the previously generated base models. If this condition is satisfied, then we calculate a new distribution D_2 over the training examples as follows.

Examples that were correctly classified by h_1 have their weights multiplied by $1/(2(1-\varepsilon_1))$.

Examples that were misclassified by h_1 have their weights multiplied by $1/(2\varepsilon_1)$. Because of the condition $\varepsilon_1 < 1/2$, correctly classified examples have their weights reduced and misclassified examples have their weights increased. Precisely, examples that h_1 misclassified have their total weight increased to 1/2 under D_2 and examples that h_1 correctly classified have their total weight reduced to 1/2 under D_2. We then go into the next iteration of the loop to construct base model h_2 using the training set and the new distribution D_2. The point is that the next base model will be generated by a weak learner; therefore, at least some of the examples misclassified by the previous base model will have to be correctly classified by the current base model. Boosting forces subsequent base models to rectify the mistakes made by earlier models. The M base models are constructed in this manner.

The ensemble returned by AdaBoost is a function that takes a new example as input and returns the class that gets the maximum weighted vote over the M base models, where each base model's weight is $\log((1-\varepsilon_m)/\varepsilon_m)$, which is proportional to the base model's accuracy on the weighted training set presented to it.

AdaBoost always performs well in practice. However, AdaBoost performs poorly when the training data is noisy.

3.2.6.2 Bagging Algorithm

Every Bootstrap AGGregatING (bagging) creates multiple bootstrap training sets from the original training set and uses each of them to generate a classifier for inclusion in the ensemble. The procedure for bagging and sampling with replacement are discussed in the following lines.

Diversity of classifiers in bagging is obtained by using bootstrapped replicas of the training data. That is, different training data subsets are randomly drawn—with replacement—from the entire training data set. Each training data subset is used to train a different classifier of the same type. Individual classifiers are then combined by taking a simple majority vote of their decisions. For any given instance, the class chosen by the greatest number of classifiers is the ensemble decision. Since the training data sets may overlap substantially, additional measures can be used to increase diversity, such as using a sub set of the training data for training each classifier, or using relatively weak classifiers (such as decision stumps).

The following algorithm votes classifiers generated by different bootstrap samples. A bootstrap sample is generated by uniformly sampling N instances from the training set with replacement.

1. Suppose $f(x, t)$ is a classifier, producing an M-vector output with 1(one) and $M-1$ (zero), at the input point x.
2. To bag $f(x, t)$, we draw bootstrap samples $T_m = (t_{1m}, t_{2m}, \cdots, t_{Nm})$ each of size N with replacement from the training data.
3. Classify input point x to the class k with largest 'vote' in $f^k_{\text{bagging}}(x, t)$ as follows.

$$f^k_{\text{bagging}}(x, t) = \frac{1}{M} \sum_{m=1}^{M} f^k_m(x, t)$$

Here the basic idea of Bagging is to reduce the deviation of several classifiers by voting the classified results due to bootstrap resamples.

Bagged ensembles tend to improve upon their base models more if the base model learning algorithms are *unstable*—differences in their training sets tend to induce significant differences in the models. Note that decision trees are unstable, which explains why bagged decision trees often outperform individual decision trees; however, decision stumps (decision trees with only one variable test) are stable, which clarifies why bagging with decision stumps tends not to improve upon individual decision stumps.

The above procedure describes the bagging algorithm. Random forests differ from this general idea in only one way: they use a modified tree learning algorithm that selects, at each candidate split in the learning process, a random subset of the features. This process is sometimes called *feature bagging*. The reason for doing this is the correlation of the trees in an ordinary bootstrap sample: if one or a few features are very strong predictors for the response variable (target output), these features will be selected in many of the trees, causing them to become correlated.

3.2.6.3 Blending

Blending (also known as stacking) is the easiest and intuitive way to combine different models. Blending involves taking the predictions of several composite models and including them in a larger model, such as a second stage linear or logistic regression. Blending can be used on any type of composite models, but is more appropriate when the composite models are fewer and more complex than in boosting or bagging. If a simple linear regression is used, this is equivalent to taking a weighted average of all predictions and works simply by reducing variance. If models are combined using logistic regression, neural network or even linear regressions with interactions, then composite models are able to have multiplicative effects on one another.

3.2.7 Naive Bayes

Naive Bayes is a simple probabilistic classifier based on Bayes' rule (Good 1992; Langley et al. 1992). The method is suitable for simple filtering and uncomplicated classification when we wish a machine to do the work. The naive Bayes algorithm builds a probabilistic model by learning the conditional probabilities of each input attribute given a possible value taken by the output attribute. This model is then used to predict an output value when we are given a set of inputs. This is done by applying Bayes' rule on the conditional probability of seeing a possible output value when the attribute values in the given instance are seen together.

Before explaining the Naïve Bayes, we define the Bayes' rule.

Bayes' rule states that

$$P(A \mid B) = \frac{P(B \mid A) P(A)}{P(B)},$$

where $P(A|B)$ is defined as the probability of observing A given that B occurs. $P(A|B)$ is called posterior probability, and $P(B|A)$, $P(A)$ and $P(B)$ are called prior probabilities. Bayes' theorem gives a relationship between the posterior probability and the prior probability. It allows one to find the probability of observing A given B when the individual probabilities of A and B are known, and the probability of observing B given A is also known.

The naive Bayes algorithm utilizes a set of training examples to classify a new instance given to it using the Bayesian approach. For example, the Bayes rule is applied to find the probability of observing each output class given the input attributes and the class that has the highest probability is assigned to the instance. The probability values used are acquired from the counts of attribute values seen in the training set.

The probability of the output attribute taking a value v_j when the given input attribute values are seen together is given by

$$P\left(v_j \mid a, b\right).$$

This probability value as such is difficult to calculate. By applying Bayes theorem on this equation, we get

$$P\left(v_j \mid a, b\right) = \frac{P\left(a, b \mid v_j\right) P\left(v_j\right)}{P\left(a, b\right)} = P\left(a, b \mid v_j\right) P\left(v_j\right),$$

where $P(v_j)$ is the probability of observing v_j as the output value, and $P(a,b|v_j)$ is the probability of observing input attribute values a, b together when output value is v_j. But if the number of input attributes $(a, b, c, d, ...)$ is large, then we likely will not have enough data to estimate the probability $P(a, b, c, d, ...|v_j)$.

The naive Bayes algorithm solves this problem by using the hypothesis of conditional independence for all the input attributes given the value for the output. So, it assumes that the values taken by an attribute are not dependent on the values of other attributes in the instance for any given output. By applying the conditional independence assumption, the probability of observing an output value for the inputs can be obtained by multiplying the probabilities of individual inputs given the output value. The probability value $P(a, b \mid v_j)$ can then be simplified as

$$P\left(a, b \mid v_j\right) = P\left(a \mid v_j\right) P\left(b \mid v_j\right),$$

where $P(a \mid v_j)$ is the probability of observing the value a for an attribute when output value is v_j. Thus the probability of an output value v_j to be assigned for the given input attributes is

$$P\left(v_j \mid a, b\right) = P\left(v_j\right) P\left(a \mid v_j\right) P\left(b \mid v_j\right).$$

Learning in the naive Bayes algorithm involves finding the probabilities of $P(v_j)$ and $P(a_i|v_j)$ for all possible values taken by the input and output attributes based on the training set provided. $P(v_j)$ is obtained from the ratio of the number of times the value v_j is seen for the output attribute to the total number of instances in the training set. For an attribute at position i with value a_i, the probability $P(a_i|v_j)$ is obtained from the number of times a_i is seen in the training set when the output value is v_j.

The naive Bayes algorithm entails all attributes in the instance to be discrete. Continuous valued attributes have to be discretized before they can be used. Missing values for an attribute are not permitted, as they can lead to problems while calculating the probability values for that attribute. A general approach to handling missing values is to substitute them with a default value for that attribute.

3.2.8 Bayesian Belief Networks

A Bayesian belief network is a directed acyclic graphical network model. It reflects the states of some part of a world that is being modelled and describes how those states are related by probabilities. The model could be of your office, for example, or your car, an ecosystem or a stock market Bayesian Belief Networks (or Bayes Nets) present the idea of applying conditional independence on a certain number of inputs rather than on all of them. This notion avoids the global assumption of conditional independence while maintaining some amount of conditional independence among the inputs.

A Bayesian belief network (Pearl 1988) provides the joint probability distribution for a set of attributes. Each attribute in the instance is represented in the network in the form of a node. In the network, a directed connection from node X to node Y is made when X is a parent of Y, which means that there is a dependence relation of Y on X, or in other words, X has an influence on Y. Thus, in this network, an attribute at a node is conditionally independent of its non-dependents in the network given the state of its parent nodes. These influences are represented by conditional probabilities, which gives the probability of a value at a node that is conditional on the value of its parents. These probability values for a node are arranged in a tabular form called a Conditional Probability Table (CPT). In the case of nodes with no parents, the CPT gives the distribution of the attribute at that node.

While a node is connected to a set of nodes that are one step above in the hierarchy, these parent nodes have an influence on its behaviour. This node is not affected by other nodes present in the given pool of nodes, meaning the node is conditionally independent of all non-parent nodes when given its parents. The nodes that are more than one step above in hierarchy, that is, the parents of parents of a node, are not considered as directly influencing the node, as they affect the nodes that are parents to the node in question and thus indirectly influence it. As a result, the parents are considered for calculating the joint probability, as only the direct parents of a node influence the conditional probabilities at this node. Using conditional independence between nodes, the joint probability for a set of attribute values $y_1, y_2,..., y_n$ represented by the nodes $Y_1, Y_2, \ldots, Y_n$ is given by

$$P(y_1, \ldots, y_n) = \prod_{i=1}^{n} P(y_i \mid \text{Parents}(Y_i)),$$

where Parents (Y_i) are the immediate parents of node Y_i. The probability values can be obtained directly from the CPTs associated with the node.

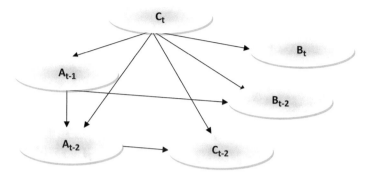

Fig. 3.7 A Bayesian network to predict temperature

A Bayesian network (Friedman et al. 1997) needs both input and output attributes to be discrete. A simple Bayesian network for predicting value (e.g., temperature), using only a few of the input instances, is shown in Fig. 3.7. Each node in the tree is associated with a CPT. For example, the CPT for the node A_{t-2} will contain the probability of each value taken by it when all possible values for A_{t-1} and C_t (i.e., its parents) are seen together. For a given instance, the Bayesian network can be used to determine the probability distribution of the target class by multiplying all the individual probabilities of values taken up by the individual nodes. The class value with the highest probability is selected. The probability of a class value taken by the output attribute C_t for the given input attributes, using parental information of nodes from the Bayesian network in Fig. 3.7 is:

$$P\left(C_t \mid A_{t-1},\, A_{t-2},\, B_t,\, B_{t-2},\, C_{t-2}\right) =$$
$$P\left(C_t\right) * P\left(A_{t-1} \mid C_t\right) * P\left(A_{t-2} \mid A_{t-1}, C_t\right) *$$
$$P\left(B_t \mid C_t\right) *$$
$$P\left(B_{t-2} \mid A_{t-1}, C_t\right) * P\left(C_{t-2} \mid A_{t-2}, C_t\right).$$

Learning in Bayes' Nets from a given training set involves finding the best performing network structure and calculating CPTs. To build the network structure, we start by assigning each attribute a node. Learning the network connections involves moving through the set of possible connections and finding the accuracy of the network for the given training set.

Bayes nets are mostly rather robust to imperfect knowledge. Often the combination of several constituents of imperfect knowledge can allow us to make remarkably strong conclusions.

3.2.9   Support Vector Machine

Support vector machine (SVM) is a classification and regression prediction tool that uses machine learning theory to maximize predictive accuracy while automatically

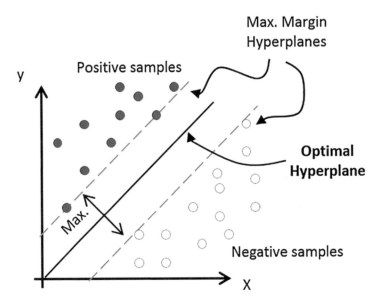

Fig. 3.8 Data classification

avoiding overfit to the data. Support vector machines (sometimes called support vector networks) can be defined as systems that use hypothesis space of linear functions in a high dimensional feature space, trained with a learning algorithm from optimization theory that implements a learning bias derived from statistical learning theory. The foundations of SVM have been developed by (Cortès and Vapnik 1995), and have gained popularity due to many promising features such as better empirical performance.

Although the SVM can be applied to various optimization problems such as regression, the classic problem is that of data classification. We can use this tool when our classification needs are straightforward. The basic idea is shown in Fig. 3.8. The data points (samples) are identified as being positive or negative, and the problem is to find a hyperplane that separates the data points by a maximal margin.

The above figure only shows the two-dimensional case where the data points are linearly separable. The problem to be solved is the following:

$$\min_{\vec{w},b} \frac{1}{2} \|w\|,$$

$$s.t \quad y_i = +1 \Rightarrow \vec{w} \cdot \vec{x}_i + b \geq +1$$
$$y_i = -1 \Rightarrow \vec{w} \cdot \vec{x}_i - b \leq -1$$

$$s.t \quad y_i \left(\vec{w} \cdot \vec{x}_i + b \right) \geq 1, \quad \forall i$$

The identification of the each data point x_i is y_i, which can take a value of $+1$ or -1 (representing positive or negative, respectively). The solution hyperplane is the following:

$$u = \overrightarrow{w} \cdot \overrightarrow{x} + b$$

The scalar b is also termed the bias.

A standard method to solve this problem is to apply the theory of Lagrange to convert it to a dual Lagrangian problem. The dual problem is the following:

$$\min_{\alpha} \Psi\left(\overrightarrow{\alpha}\right) = \min_{\alpha} \frac{1}{2} \sum_{i=1}^{N} \sum_{j=1}^{N} y_i y_j \left(\overrightarrow{x}_i \cdot \overrightarrow{x}_j\right) \alpha_i \alpha_j - \sum_{i=1}^{N} \alpha_i$$

$$\sum_{i=1}^{N} \alpha_i y_i = 0$$

$$\alpha_i \geq 0, \quad \forall i$$

The variables α_i are the Lagrangian multipliers for corresponding data point x_i.

Classification in SVM is an example of supervised learning. Known labels help indicate whether the system is performing in a right way or not. This information points to a desired response, validating the accuracy of the system, or is used to help the system learn to act correctly. A step in SVM classification involves identification of which are intimately connected to the known classes. This is called feature selection or feature extraction. Feature selection and SVM classification together have a use even when prediction of unknown samples is not necessary. They can be used to identify key sets that are involved in whatever processes distinguish the classes.

SVMs can also be applied to regression problems by the introduction of an alternative loss function. The loss function must be modified to include a distance measure. The regression can be linear and non-linear. Similarly to classification problems, a non-linear model is usually required to adequately model data. In the same manner as the non-linear SVC approach, a non-linear mapping can be used to map the data into a high dimensional feature space where linear regression is performed. The kernel approach is again employed to address the curse of dimensionality. In the regression method, there are considerations based on prior knowledge of the problem and the distribution of noise.

Support vector machines act as one of the effective approaches to data modelling. They combine generalization control as a technique to control dimensionality. The kernel mapping provides a common base for most of the commonly employed model architectures, enabling comparisons to be performed. The major strength of SVM is that the training is relatively easy. There is no local optimal, unlike in neural networks. SVM scale relatively well to high dimensional data and the trade-off between classifier complexity and error can be controlled explicitly. The weakness includes the need for a good kernel function.

3.3 Unsupervised Learning

In unsupervised learning, we are dealing with unlabelled data or data of *unknown structure*. Using unsupervised learning techniques, we are able to explore the structure of our data to extract meaningful information without the guidance of a known outcome variable or reward function.

With unsupervised learning, it is possible to learn larger and more complex models than with supervised learning. This is because in supervised learning, one is trying to find the connection between two sets of observations. The difficulty of the learning task increases exponentially in the number of steps between the two sets and that is why supervised learning cannot, in practice, learn models with deep hierarchies.

In unsupervised learning, the learning can proceed hierarchically from the observations into ever more abstract levels of representation. Each additional hierarchy needs to learn only one step, and therefore the learning time increases linearly in the number of levels in the model hierarchy.

For example, you can ask a panel of oncologists to review a set of breast images and classify them as possibly malignant (or not), but the classification is not a part of the raw source data. Unsupervised learning techniques help the analyst identify data-driven patterns which may warrant further investigation.

Clustering is an exploratory unsupervised technique that allows us to organise a pile of information into meaningful subgroups (*clusters*) without having any prior knowledge of their group memberships. Each cluster that may arise during the analysis defines a group of objects which share a certain degree of similarity but are more dissimilar to objects in other clusters, which is why clustering is sometimes also called 'unsupervised classification'. Clustering is a popular technique for structuring information and deriving meaningful relationships among data; for example, it allows marketers to discover customer groups based on their interests in order to develop distinct marketing programs.

3.3.1 Apriori Algorithm

Apriori is an algorithmic solution for frequent itemset mining and association rule learning over transactional databases. In order to perform predictive analysis, it is useful to discover interesting patterns in the given data set that serves as the basis for estimating future trends. This refers to the discovery of attribute-value associations that frequently occur together within a given data set.

Let us suppose that $A = \{l_1, l_2, \ldots l_m\}$ is a set of items and T is a set of transactions, where each transaction t is a set of items. Thus, t is a subset of A.

A transaction t is said to support an item l_i, if l_i is present in t. t is said to support a subset of items $X \subseteq A$, if t supports each item l_i in X. An itemset $X \subseteq A$ has a support in T, denoted by $s(X)_T$, if a certain percentage of transactions in T supports X.

Support can be defined as a fractional support, denoting the proportion of transactions supporting X in T. It can also be discussed in terms of the absolute number of transactions supporting X in T. For ease, we will assume the support to be %-support. We will often drop the subscript T in the expression $s(X)_T$, when T is apparently implied.

For a given transaction database T, an *association rule* is an expression of the form $X \Rightarrow Y$, where X and Y are subsets of A. The rule $X \Rightarrow Y$ holds with *confidence* τ, if $\tau\%$ of transactions in D that support X also support Y. The rule $X \Rightarrow Y$ has *support* σ in the transaction set T if $\sigma\%$ of the transactions in T support $X \cup Y$.

Every algorithm for finding association rules assumes that the basic database is very large, requiring multiple passes over the database. The idea is to achieve all rules fulfilling pre-specified frequency and accuracy criteria. In real data sets, there are relatively few frequent sets. For instance, generally customers will buy a small subset of the entire collection of products. If data sets are large enough, it will not fit into main memory. Therefore, one can aim at the procedures that read the data as few times as possible. Algorithms that obtain association rules from data usually divide the task into two parts—first, find the frequent itemsets, and then form the rules from the frequent sets. That means the problem of mining association rules can be divided into two subproblems:

- Find itemsets whose support is greater than the user-specified minimum support, σ. Those itemsets are referred to as frequent itemsets.
- Use the frequent itemsets to produce the desired rules. In general if $ABCD$ and AB are frequent itemsets, then we can determine if the rule $AB \Rightarrow CD$ holds by checking the following inequality

$$\frac{s(\{A, B, C, D\})}{s(\{A, B\})} \geq \tau$$

where $s(X)$ is the support of X in T.

Let T be the transaction database and σ be the user-specified minimum support. An itemset $X \subseteq A$ is said to be a *frequent itemset* in T with respect to σ, if

$$s(X) \geq \sigma$$

A frequent set is a *maximal frequent set* if it is a frequent set and no superset of this is a frequent set.

Real World Case 5: Learning Association Rules in Marketing
Shopping malls use association rules to place the items next to each other so that users purchase more items. With the help of machine learning and data mining, companies can now think through how and what people buy, and therefore lay out stores more efficiently. A popular case study, the 'Beer–Diapers Wal-Mart- story', is a common example. Wall-Mart studied their data and found that on Friday afternoons, young males who buy diapers also tend to buy beer. Hence, the manager decided to place beer next to diapers and the beer sales went up.

An itemset is a *border set* if it is not a frequent set, but all its proper subsets are frequent sets. Furthermore, if X is an itemset that is not frequent, then it should have a subset that is a border set.

If we know the set of all maximal frequent sets of a given T with respect to a σ, then we can find the set of all frequent sets without any extra check of the database. Thus, the set of all maximal frequent sets can act as a compact representation of the set of all frequent sets. However, if we need the frequent sets together with their relevant support values in T, then we have to make one more database pass to derive the support values, as the set of all maximal frequent sets is known.

A maximal frequent set may or may not be a proper subset of a border set. It is likely that a proper subset of a border set, of cardinality one less than the border set, is not necessarily maximal. Hence, it is difficult to determine an exact relationship between the set of maximal frequent sets and the set of border sets. However, the set of all border sets and the maximal frequent sets, which are not proper subsets of any of the border sets, together propose enhanced representation of the set of frequent sets.

Agrawal and Srikant developed the *Apriori algorithm* in 1994 (Agrawal and Srikant 1994). It is also called the *level-wise algorithm.* It is the most accepted algorithm for finding all frequent sets. It makes use of the downward closure property. The algorithm is a bottom-up search, progressing upward level-wise in the lattice. But, the interesting fact about this method is that before reading the database at every level, it prunes many of the sets which are unlikely to be frequent sets. In general, the algorithm works as follows:

- The first pass of the algorithm simply counts item occurrences to determine the frequent itemsets.
- A subsequent pass, say pass k, consists of two phases.

  (a) The frequent itemsets L_{k-1} found in the $(k-1)^{th}$ pass are used to generate the candidate itemsets C_k, using the a priori candidate-generation procedure described below.
  (b) The database is scanned and the support of candidates in C_k is counted.

For fast counting, one needs to efficiently determine the candidates in C_k contained in a given transaction t. The set of candidate itemsets is subjected to a pruning process to ensure that all the subsets of the candidate sets are already known to be frequent itemsets. The candidate generation process and the pruning process are the very important parts of this algorithm, and are described below.

Let us consider L_{k-1}, the set of all frequent $(k-1)$-itemsets. We wish to create a superset of the set of all frequent k-itemsets. The insight behind the a priori candidate-generation procedure is that if an itemset X has minimum support, then consider all subsets of X.

A priori candidate generation procedure is given below:

$C_k = \{\}$
for all itemsets $l_1 \in L_{k-1}$ do
for all itemsets $l_2 \in L_{k-1}$ do
if $l_1[1] = l_2[1] \wedge l_1[2] = l_2[2] \wedge \cdots \wedge l_1[k-1] < l_2[k-2]$
then $c = l_1[1], l_1[2], \ldots, l_1[k-1], l_2[k-1]$
$C_k = C_k \cup \{c\}$

The pruning algorithm described below prunes some candidate sets, which do not meet the second criterion.

Prune (C_k)
for all $c \in C_k$
for all $(k-1)$ − subsets d of c do
if $d \notin L_{k-1}$

then $C_k = C_k \setminus \{c\}$

The pruning step eliminates the extensions of $(k-1)$-itemsets that are not found to be frequent, from being considered for counting support.

The a priori frequent itemset discovery algorithm uses these two functions at every iteration. It moves upward in the lattice starting from level 1 to level k, where no candidate set remains after pruning.

Now the Apriori algorithm is given as:

Initialize $k := 1$, C_1 all the 1-itemsets;
read the database to count the support of C_1, to determine L_1
$L_1 := \{frequent\ 1 - itemsets\}$;
$k := 2$; //k represents the pass number
while $(L_{k-1} \neq \{\})$ do
begin
$C_k :=$ gen_candidate_itemsets with the given L_{k-1}
prune (C_k)
for all transactions $t \in T$ do

increment the count of all candidates in C_k that are contained in
$L_k :=$ All candidates in C_k with minimum support;
$k := k + 1$
end

We have understood that searching frequent patterns in transactional databases is considered to be one of the key data analysis problems and Apriori is one of the prominent algorithms for this task. Developing fast and efficient algorithms that can handle large volumes of data becomes a challenging task. To tackle the challenge, one can formulate a parallel implementation of Apriori algorithm (a kind of MapReduce job). The `map` function performs the procedure of counting each occurrence of potential candidates of size k and thus the map stage realizes the occurrences counting for all the potential candidates in a parallel way. Then, the `reduce` function performs the procedure of summing the occurrence counts. For each round of the iteration, such a job can be carried out to implement occurrence computing for potential candidates of size k.

3.3.2 k-Means Algorithm

The k-means method of cluster detection is the most commonly used in practice. The method is useful when we have a substantial number of groups to classify. It has many variations, but the form described here was the one first published by MacQueen (1967). This algorithm has an input of predefined number of clusters, which is called k. 'Means' stands for an average: the average location of all the members of a single cluster.

Let us assume that the data is represented as a relational table, with each object representing an object and each column representing a column. The value of every attribute from the table of data we are analysing represents a distance from the origin along the attribute axes. Furthermore, to use this geometry, the values in the data set must all be numeric. If they are categorical, then they should be normalized in order to allow adequate results of the overall distances in a multi-attribute space. The k-means algorithm is a straightforward iterative procedure, in which a vital notion is that of *centroid*. A centroid is a point in the space of objects which represents an average position of a single cluster. The coordinates of this point are the averages of attribute values of all objects that belong to the cluster. The iterative process, of redefining centroids and reassigning data objects to clusters, needs a small number of iterations to converge. The simple stepwise description of k-means can be given as:

Step 1. Randomly select k points to be the starting points for the centroids of the k clusters.
Step 2. Assign each object to the centroid closest to the object, forming k exclusive clusters of examples.

Step 3. Calculate new centroids of the clusters. Take the average of all attribute values of the objects belonging to the same cluster.

Step 4. Check if the cluster centroids have changed their coordinates.

If yes, repeat from the Step 2.

If no, cluster detection is finished and all objects have their cluster memberships defined.

Real World Case 6: Market Price Modelling
A standard challenge facing many automotive manufacturers is developing correct price quotations for their customers. In the locomotive industry, auto-motive subsystems that go into a car are complex assemblies of several parts. Whenever a new car is designed, some subsystems need to be changed in some way—from minor adjustments to occasionally a new design altogether. This requires an intense understanding of their subsystems, from materials, to engineering to manufacturing. Developing such a level of insight is possible with gathered data. k-Means clustering can be successfully employed to promptly identify relationships between the different underlying parts and the total cost to manufacture. Using k-means clustering, an identification of accurate pricing is possible by plotting the centroids of every attribute for the various clusters.

Now, let us discuss the partitioning technique in detail.

The purpose of clustering is to obtain subsets that are more genuine than the initial set. This means their elements are much more similar on average than the elements of the original domain. A partition $T_1, T_2, \ldots, T_k$ is represented by the centroids $z_1, z_2, \ldots, z_k$ such that

$$x \in T_i \iff \rho(x, z_i) \leq \rho(x, z_j), i, j = 1, \ldots k$$

One can see that even though no information about the classes has been used in this case, the k-means algorithm is perfectly capable of finding the three main classes.

The centroids are used for estimates of an *impurity measure* of the form

$$J(z_1, z_2, \ldots, z_p) = \frac{1}{N} \sum_{i=1}^{k} \sum_{x^{(j)} \in T_i} \rho(x^{(j)}, z_i) = \frac{1}{N} \sum_{j=1}^{N} \min_{1 \leq i \leq k} \rho(x^{(j)}, z_i)$$

The algorithms for partitioning (i.e., k-means and k-medoid) vary in the manner that they estimate the centroids. In the k-means algorithm, the mean of the real-valued observations in the cluster T_i is calculated as:

$$z_i = \frac{1}{N_i} \sum_{x^{(j)} \in T_i} x^{(j)},$$

where N_i denotes the number of data points in T_i.

We can observe an interesting property in that the k-means algorithm will not increase the function J. On the contrary, if any clusters are changed, J is reduced. As J is bounded from below, it converges and as a consequence the algorithm converges. It is also shown that the k-means will always converge to a local minimum.

There are two key steps in the algorithm: the determination of the distances between all the points and the recalculation of the centroids.

Two disadvantages of the k-means method are that the mean may not be close to any data point at all, and the data is limited to real vectors.

An alternative algorithm is the k-medoid where the centroid is chosen to be the most central element of the set.

That is, $z_i = x^{(s_i)}$

such that

$$\sum_{x^{(j)} \in T_i} \rho\left(x^{(j)}, x^{(s_i)}\right) \leq \sum_{x^{(j)} \in T_i} \rho\left(x^{(j)}, x^{(m)}\right) \text{for all } x^{(m)} \in T_i.$$

3.3.3   Dimensionality Reduction for Data Compression

Another subfield of unsupervised learning is *dimensionality reduction*. Often we are working with data of high dimensionality—each observation comes with a high number of measurements—that can present a challenge for limited storage space and the computational performance of machine learning algorithms. Unsupervised dimensionality reduction is a commonly used approach in feature pre-processing to remove noise from data, which can also degrade the predictive performance of certain algorithms, and compress the data onto a smaller dimensional subspace while retaining most of the relevant information.

There are many diverse examples of high-dimensional data sets that are difficult to process at once: videos, emails, user logs, satellite observations, and even human gene expressions. For such data, we need to throw away unnecessary and noisy dimensions and keep only the most informative ones. A classic and well-studied algorithm for reducing dimension is Principal Component Analysis (PCA), with its non-linear extension Kernel PCA (KPCA). Assuming that data is real-valued, the goal of PCA is to project input data onto a lower dimensional subspace, preserving as much variance within the data as possible.

3.4 Reinforcement Learning

Reinforcement learning is a branch of machine learning concerned with using experience gained through interacting with the world and evaluative feedback to improve a system's ability to make behavioural decisions.

In supervised learning we have assumed that there is a target output value for each input value. However, in many situations, there is less detailed information available. In extreme situations, there is only a single bit of information after a long sequence of inputs telling whether the output is right or wrong. Reinforcement learning is one method developed to deal with such situations. Reinforcement learning is a computational approach to understanding and automating goal-directed learning and decision making. It is the problem experienced by an agent that must learn behaviour through trial-and-error exchanges with a dynamic environment. There are basically two main approaches for solving reinforcement-learning problems. The first is to search in the space of behaviours to find one that performs well in the environment. This approach has been taken by work in genetic algorithms and genetic programming. The second is to use statistical techniques and dynamic programming methods to estimate the benefit of taking actions in states of the world.

Reinforcement learning has three fundamental components:

- *Agent*: the learner or the decision maker.
- *Environment*: everything the agent interacts with, i.e., everything *outside* the agent.
- *Actions*: what the agent can do.

Each action is associated with a reward. The objective is for the agent to choose actions so as to maximize the expected reward over some period of time.

Moreover, there are four constituents of a reinforcement learning system (an agent): a policy, a reward function, a value function, and a model of the environment.

A policy is the decision-making function of the agent. It is a mapping from perceived states of the environment to actions to be taken when in those states. The policy is the core of a reinforcement learning agent in the sense that it alone is sufficient to determine behaviour. In general, policies may be stochastic. It specifies what action the agent should take in any of the situations it might encounter. The other components serve only to change and improve the policy.

A reward function defines the goal in a reinforcement learning problem. Roughly speaking, it maps each perceived state (or state-action pair) of the environment to a single number, a reward, indicating the intrinsic desirability of that state. A reinforcement learning agent's sole objective is to maximize the total reward it receives in the long run. The reward function defines what the good and bad events are for the agent. The value of a state is the total amount of reward an agent can expect to accumulate over the future, starting from that state. Whereas rewards determine the immediate, intrinsic desirability of environmental states, values indicate the long-term desirability of states after taking into account the states that are likely to follow, and the rewards available in those states.

The fourth and final element of some reinforcement learning systems is a model of the environment. This is something that mimics the behaviour of the environment. For example, given a state and action, the model might predict the resultant next state and next reward. Models are used for planning, by which we mean any way of deciding on a course of action by considering possible future situations before they are actually experienced.

The term 'reinforcement learning' and some of the general characteristics of the technique have been borrowed from cognitive psychology, but in recent years it has become increasingly popular within machine learning. Reinforcement learning describes any machine learning technique in which a system learns a policy to achieve a goal through a series of trial-and-error training sessions, receiving a reward or punishment after each session, and learning from this 'reinforcement' in future sessions.

Formally, the reinforcement learning problem is as follows:

- a discrete set of agent actions, A,
- a discrete set of environment states, S, and
- a set of scalar reinforcement signals, typically $\{0, 1\}$, or the real numbers.

The goal of the agent is to find a policy π, mapping states to actions, that maximizes some long-run measure of reinforcement. We will denote the optimal policy as π^*. Various techniques exist within reinforcement learning for choosing an action to execute in a particular state. Q-learning is one of the more traditional of these techniques.

Q-learning falls within the 'model-free' branch of the reinforcement learning family, because it does not require the agent to learn a model of the world or environment (i.e., how actions lead from state to state, and how states give rewards) in order to learn an optimal policy. Instead, the agent interacts with the environment by executing actions and perceiving the results they produce. In Q-learning, each possible state-action pair is assigned a quality value, or 'Q-value' and actions are chosen by selecting the action in a particular state with the highest Q-value.

The process works as follows. The agent can be in one state of the environment at a time. In a given state $s_i \in S$, the agent selects an action $a_i \in A$ to execute according to its policy π. Executing this action puts the agent in a new state s_{i+1}, and it receives the reward r_{i+1} associated with the new state. The policy in Q-learning is given by:

$$\pi(s) = \arg\ \max_a Q(s,\ a),$$

and thus by learning an optimal function $Q^*(s,\ a)$ such that we maximize the total reward r, we can learn the optimal policy $\pi^*(s)$.

We update $Q(s,\ a)$ after taking each action so that it better approximates the optimal function $Q^*(s,\ a)$. We do this with the following equation:

$$Q(s_i,\ a_i)\ :=\ r_i + \gamma \max_{a'} Q(s_{i+1},\ a'),$$

where γ is the discount factor, $0 < \gamma < 1$.

In the first few training sessions, the agent has not yet learned even a rough approximation of the optimal Q-function, and therefore the initial actions will be random if the initial Qs are. However, due to this randomness, the danger exists that, although the Q-function may eventually converge to a solution to the problem, it may not find the *optimal* solution to the problem, because the initial path it stumbles upon may be a correct but suboptimal path to the goal. Therefore, it is necessary that every reinforcement learning agent have some policy of exploration of its environment, so that if a better solution than the one it has already found exists, it may stumble upon it.

One standard approach for arbitrating the trade-off between exploitation (using knowledge already learned) and exploration (acting randomly in the hopes of learning new knowledge) is to choose each action with a probability proportionate to its Q-value. In particular:

$$Pr\,(a_i\,|\,s) = \frac{T^{-Q(s,a_i)}}{\sum_j T^{-Q(s,a_j)}}.$$

The temperature, T, starts at a high value and is decreased as learning goes on. At the beginning, the high value of T ensures that some actions associated with low Q-values are taken—encouraging exploration. As the temperature is decreased, most actions are those associated with high Q-values—encouraging exploitation. This is the method of choosing actions we will use.

Note that the Q-learning is exploration insensitive. The Q values will converge to the optimal values, independent of how the agent behaves while the data is being collected. This means that although the exploration-exploitation issue must be addressed in Q-learning, the details of the exploration strategy will not affect the convergence of the learning algorithm. For these reasons, Q-learning is the most popular and seems to be the most effective model-free algorithm for learning from delayed reinforcement. It does not, however, address any of the issues involved in generalizing over large state and/or action spaces. In addition, it may converge quite slowly to a good policy.

Like Q-learning, P-learning is a technique for choosing actions in a particular state. In fact, P-learning is nearly identical to Q-learning, except that rather than remembering a real-valued scalar to represent the quality of each state-action pair, the agent only remembers a 1 if the action in the given state is optimal, and a 0 if the action is non-optimal. Formally:

if $a \in \pi * (s)$ then $P\,(s,\ a) = 1$ else $P\,(s,\ a) = 0$.

In practice, the P function is computed from the Q function (which must also be learned), but it can be stored more compactly and used later more easily. The P function is expressed in terms of the Q function as follows:

if $a \in \arg\max_a Q(s, a)$ then $P\,(s,\ a) = 1$ else

$$P\,(s,\ a) = 0.$$

We can also incorporate exploration into P-learning through the analogous equation to the one for choosing actions in Q-learning:

$$Pr\left(a_i \mid s\right) = T^{-P(s, \, a_i)}/\mathbf{S}_j T^{-P(s, \, a_j)}.$$

3.4.1 Markov Decision Process

Reinforcement learning has strong connections with optimal control, statistics and operational research. Markov decision processes (MDP) are popular models used in reinforcement learning. MDP assume that the state of the environment is perfectly observed by the agent. When this is not the case, one can use a more general model called partially observable MDP to find the policy that resolves the state uncertainty while maximizing the long-term reward.

Markov decision processes (Bellman 1957) are an important tool for modelling and solving sequential decision making problems. MDP provide a mathematical framework for modelling decision making in situations where outcomes are partly random and partly under the control of the decision maker. They originated in the study of stochastic optimal control in the 1950s and have remained of key importance in that area ever since. Today, MDPs are used in a variety of areas, including robotics, automated control, planning, economics and manufacturing.

A reinforcement learning task that satisfies the Markov property is called a *Markov decision process*, or *MDP*. If the state and action spaces are finite, then it is called a *finite Markov decision process (finite MDP)*.

A finite MDP is defined by its state and action sets and by the one-step dynamics of the environment. Given any state and action, S and a, the probability of each possible next state, s', is

$$P_{ss'}^{a} = \Pr\left\{s_{t+1} = s' \mid s_t = s, a_t = a\right\}.$$

These quantities are called *transition probabilities*. Similarly, given any current state and action, S and a, together with any next state, s', the expected value of the next reward is

$$R_{ss'}^{a} = E\left\{r_{t+1} \mid s_t = s, a_t = a, s_{t+1} = s'\right\}.$$

These quantities, $P_{ss'}^{a}$, and $R_{ss'}^{a}$, completely specify the most important aspects of the dynamics of a finite MDP.

While solving MDP, we always try to obtain the optimal policy, which is defined as the policy with an expected return greater than or equal to all other policies for all states. There can be more than one optimal policy. The MDP framework is abstract, flexible, and provides the tools needed for the solution of many important real world problems. The flexibility of the framework allows it not only to be applied to many different problems, but also in many different ways. For example, the time steps can refer to arbitrary successive stages of decision making and acting. The actions can be any decisions we want to learn how to make and the state can contain anything that might be useful in making them.

3.5   Case Study: Using Machine Learning for Marketing Campaign

A well-known furniture company wants to expand its marketing strategy to display product advertisements online. A key challenge is to create a campaign that gets *right* clicks from customers. The company manager has planned different advertisement campaigns to attract the rich customers. After 2 months, the analytics team told the manager which campaigns brought in the highest average revenue and even the number of top customers; however, the manager is interested to know what elements are attracting the top shoppers in order to enhance marketing strategy further.

At this juncture, the analytics team delves into machine learning techniques to solve problems by finding patterns that they cannot see themselves.

For instance, considering the target campaign, they wanted to offer special discounts to newly married couples on the items they needed as a new family — with anticipation of turning them into an ultimate, loyal customer. But the main issue is how to find them? So, they started identifying buying habits of someone who had recently married. They used a machine learning to detect post-marriage buying patterns. The marketing team was then able to make sure that these customers received direct mail with the special offers.

Steps for developing learning rules are:

1. *Find features*: Consider a real world problem and map it in spreadsheet, the columns are the different 'features' of a campaign. The rows are the data points. What features of each advertisement led to the purchase? Which photo (e.g., logo, product) did they see? Which platform (such as Google) did they click from?
2. *Identify outcomes*: Here analysts need to have both an appropriate clear outcome and a negative outcome. This helps the machine to find the pattern that leads, most often, to the right outcome. For example, a desirable result is a spending of 500 Euros or more, and a negative result is under 500 Euro. So, a simple rule is: if the customer spent more than 500 Euro then we use 'true' and if not, false.
3. *Collect the right data*: Here, the analytics team has to tag each link in the advertisement with the proper URL variable so that when a potential shopper clicks, they know the platform, the advertisement, and the photo that brought them to the site. Next, they have to combine the data with the features. (But what if they do not have the right data? Think!)
4. *Select proper machine learning algorithm/tool*: We now know that there are several machine learning algorithms. Let us recall the last half of Sect. 3.1, and

the crucial question of selecting an appropriate machine learning algorithm in the context of your requirements. Every machine learning algorithm has its own use case, which can produce some very complex models to help you predict the future. Good features allow a simple model to beat a complex model. In this case study, you can choose the model which obviously tells you what is effective from a machine learning perspective: a decision tree.

5. *Split your data*: The splitting is essential for making two sets: one set of data for learning and another for testing. Typically the learning data is much larger than the testing. For instance, the analytics team used 1,200 items for learning and then test it on 300 items. The ultimate aim is to check whether the model that is built by the machine from the learning data essentially works on the testing data.

6. *Run the machine learning algorithm*: For instance, employ decision tree tool/package in R.

7. *Assess the results*: The analyst has to assess output file with some interpretation.

8. *Finally, take action*: A machine learning marketing system would follow rules automatically and keep running the tree with new data in order to improve results and improve predictive results. This is where machine learning gets really interesting, as it culminates in a system which changes and improves itself over time.

From these steps, you can see the preparation necessary to use machine learning with a marketing program so that you can take steps towards computer-assisted marketing automation.

3.6   Exercises

1. Suppose we have a large training set. Name a drawback when using a k nearest neighbour during testing.

2. What is a decision tree? In the ID3 algorithm, what is the expected information gain, and how is it used? What is the gain ratio, and what is the advantage of using the gain ratio over using the expected information gain? Describe a strategy that can be used to avoid overfitting in decision trees.

3. Develop a table of training examples in some domain, such as classifying animals by species, and trace the construction of a decision tree by the ID3 algorithm.

4. In most learning algorithms, the computation time required for training is large and the time required to apply the classifier is short. The reverse is true of the nearest neighbour algorithm. How can the computation time required for queries be reduced?

5. Explain the principle of the *k*-Means algorithm. Describe situations where *k*-Means should be used or not used.

6. Explain the principle of AdaBoost with simple stumps. Give an example of a data set that can be learned by AdaBoost, and one that is not optimal.

7. (**Project**) Explore how AdaBoost can be used for face detection. How is it trained? What are the features? How is it applied?
8. What kind of problem can be solved with reinforcement learning algorithms? What are the advantages and the disadvantages of this technique?
9. (**Project**) Implement the Apriori algorithm in a language of your choice and run it on any examples of your choice.
10. A learning theory may be divided into the following parts:

   1. A hypothesis space.
   2. A representation for hypotheses.
   3. A preference criterion over hypotheses, independent of the data.
   4. A measure of how well a given hypothesis fits given data.
   5. A search strategy to find a good hypothesis for a given data set.

For the machine learning method of your choice, explain what each of these is.

References

Agrawal, R., & Srikant, R. (1994). *Fast algorithms for mining association rules in large databases.* In Proceedings of the 20th international conference on very large data bases. Santiago: VLDB.

Akerkar, R., & Lingras, P. (2007). *Building an intelligent web: Theory & practice.* Sudbury: Jones & Bartlett Publisher.

Bellman, R. (1957). A Markovian decision process. *Journal of Mathematics and Mechanics, 6,* 679–684.

Breiman, L. (2001). Random forests. *Machine Learning, 45*(1), 5–32.

Breiman, L., Friedman, J., Olshen, R., & Stone, C. (1984). *Classification and regression trees.* Belmont: Wadsworth.

Cortes, C., & Vapnik, V. (1995). Support-vector networks. *Machine Learning, 20*(3), 273–297.

Friedman, N., Geiger, D., & Goldszmidt, M. (1997). Bayesian network classifiers. *Machine Learning, 29,* 131–163.

Good, I. (1992). *The estimation of probabilities: An essay on modern Bayesian methods.* Boston: M.I.T. Press.

Langley, P., Iba, W., & Thompson, K. (1992). *An analysis of Bayesian classifiers.* Proceedings of the tenth national conference on artificial intelligence. s.l.: AAAI Press.

MacQueen, J. (1967). *Some methods for classification and analysis of multivariate observations.* In Proceedings of 5th Berkeley symposium on mathematical statistics and probability. s.l.: University of California Press.

Mitchell, T. (1997). *Machine learning.* s.l.: McGraw Hill.

Nilsson, N. (1996). *Introduction to machine learning.* s.l.:Department of Computer Science, Stanford University.

Oza, N. (2004). *Ensemble data mining methods.* NASA.

Pearl, J. (1988). *Probabilistic reasoning in intelligent systems.* San Francisco: Morgan Kaufman.

Quinlan, R. (1986). Induction of decision trees. *Machine Learning, 1,* 81–106.

Quinlan, R. (1993). *C4.5: Programs for machine learning.* San Francisco: Morgan Kaufmann.

Witten, I., & Frank, E. (2005). *Data mining: Practical machine learning tools and techniques* (2nd ed.). San Francisco: Morgan Kaufmann.

Chapter 4
Fuzzy Logic

4.1 Introduction

Set is defined as a collection of entities that share common characteristics. From the formal definition of the set, it can be easily determined whether an entity can be member of the set or not. Classically, when an entity satisfies the definition of the set completely, then the entity is a member of the set. Such membership is certain in nature and it is very clear that an entity either belongs to the set or not. There is no intermediate situation. Thus, the classical sets handle bi-state situations and sets membership results in either 'true' or 'false' status only. These types of sets are also known as crisp sets. In other words, a crisp set always has a pre-defined boundary associated with it. A member must fall within the boundary to become a valid member of the set. An example of such classical set is the number of students in a class, 'Student'. Students who have enrolled themselves for the class by paying fees and following rules are the valid members of the class 'Student'. The class 'Student' is crisp, finite and non-negative. Here are some types of crisp sets.

- *Finite set example*: Set of all non-negative integers less than 10. Clearly, the set contains 0, 1, 2, 3, 4, 5, 6, 7, 8 and 9.
- *Infinite set example*: Set of all integers less than 10. There are many numbers that are less than 10. The numbers starts with 9, 8, 7, 6, 5, 4, 3, 2, 1, 0, −1, −2, and so on. The set is infinite, but it is possible to determine if a number is member of a set of not.
- *Empty sets*: Set of live fishes on earth. Obviously, the set has no members, as a fish cannot live on earth; hence the set is known as an empty set.

Unlike the crisp sets or the classical sets, fuzzy sets allow an entity to possess a partial membership in the sets. Entities that strictly and completely satisfy a fuzzy set definition, are obviously part of the fuzzy set, but the entities that do not strictly follow the definition, are also members of the fuzzy set to some extent, with partial truth value. Entities that completely and strictly satisfy the membership

© Springer International Publishing Switzerland 2016
R. Akerkar, P.S. Sajja, *Intelligent Techniques for Data Science*,
DOI 10.1007/978-3-319-29206-9_4

criteria are provided truth value 1. Entities that do not completely and strictly satisfy the membership criteria are provided truth value 0. These are two extreme cases of belongingness; 1 means completely belonging and 0 means complete non-belonging. The entities that partially belong to the fuzzy set are typically given values between 0 and 1. These values are known as membership values. Considering this, we can state that the fuzzy set is a superset of the crisp set for given variable / domain. It is clear that a fuzzy set does not have a sharp, but rather an open boundary, as it allows partial membership to the set.

Fuzzy logic uses fuzzy sets at as its base. Lofti Zadeh (1965), who is known as the father of fuzzy logic, has claimed that many real life sets are defined with an open or no boundary. Instead of sticking with only 'true' and 'false' values of a variable, he introduced a gradual improvement from false to true values. That is, instead of only two truth values, he has introduced the concept of many truth values between these extremes. This work also documents the fuzzy set theory and fuzzy logic as an extension of the fuzzy set theory. He has identified some characteristics of typical fuzzy sets as follows:

• Any logical system can be fuzzified.
• Fuzzy logic provides partial membership to a set, while for a fuzzy set/logic every belonging is a matter of degree.
• Reasoning in the fuzzy approach is viewed as an approximate reasoning and considered as an extended case of general reasoning.
• Knowledge is interpreted as a collection of elastic or, equivalently, fuzzy constraint on a collection of variables.
• Inference is viewed as a process of propagating elastic constraints.

It is true that we commonly and subconsciously classify things into classes with open boundaries. Terms like 'luxury car', 'difficult problem', and 'young person' are used freely in our day-to-day endovers (operations). For people, it is easy to understand their meaning and significance, but machines cannot handle such words without the help of fuzzy sets and fuzzy logic. A mapping function is required to help in converting fuzzy values into their equivalent crisp values (fuzzification) and vice versa (defuzzification). It is the proper fuzzy membership function that helps such user-friendly interaction by defuzzifying and fuzzifying such values.

Let us see an example to clarify the concept of fuzzy set and crisp set. Consider a set of heighted (tall) people. If a person has a height of 5 ft 6 in, then a person is considered to be tall. This is a crisp definition of a set of heighted people. In this case, if a person's height is 1 in less (or even lesser height), he seem to be a heighted person at first sight, but is not eligible to be a part of the set of heighted people. Further, the person with a height of 3 ft 5 in (who is considerably short) is treated equal to a person with a height of 5 ft 4 in. Neither fit into the class of heighted people according to the rigid definition of the set. Fuzzy sets solve such problems by assigning partial belongingness values to candidate persons. Here, the

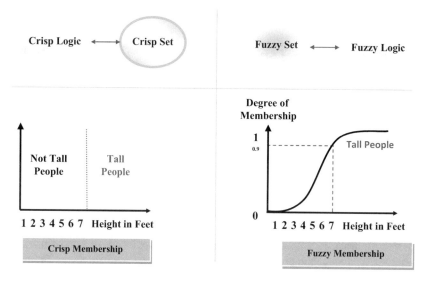

Fig. 4.1 Crisp and fuzzy set example: set of heighted people

person with a height of 5 ft 4 in is also a member of a set, but not completely, with a membership degree nearer to 1 (say 0.9). Similarly, the person who is 3 ft 5 in is also a member of the set of heighted people, with a membership degree nearer to 0 (say 0.2). The values nearer to 1 show stronger belonging and membership values nearer to 0 show poorer belonging to the fuzzy set. This situation is shown in Fig. 4.1.

It is to be noted that a member of a fuzzy set can also be a member of another fuzzy set. That means the person with a height of 5 ft 4 in is a member of the fuzzy set of heighted people, with a membership degree of 0.9; and simultaneously the person is also member of the fuzzy set defining short people, with a membership value of 0.1. In this sense, membership of a fuzzy set is not complete. A kind of vagueness is there, as the fuzzy sets do not have sharp boundaries.

Here is the formal definition of a fuzzy set.

A fuzzy set A in the universe of discourse U (containing all elements that can be considered) is defined as a set of order pairs, each containing an element and its membership degree as given below.

$$A = \{(X, \mu_A (X)), \text{ where } X \in U \text{ and } \mu_A (X) \in [0, 1]\}$$

The fuzzy set of heighted people can be formally defined as follows.

Let us name the set as F_Tall, indicating it is a set of tall people having good height.

$$F\_Tall = \{(X, \mu (X)), \text{ where } 0 <= \mu (X) <= 1\}$$

4.2 Fuzzy Membership Functions

The graded membership μ of a person X to determine whether the person belongs to set of tall people (T) is defined as a formal membership function, as shown below:

$\mu_T(X) = 0$ if height of X is $<=$ 4.5 feet
$\mu_T(X) = (\text{height of } X - 4.5)$ if height of X is $>$ 4.5 feet and $<=$ 5.5 feet
$\mu_T(X) = 1$ elsewhere

The membership function helps in converting crisp data to fuzzy values and vice versa. That is, once you know the membership function and the membership degree, you can find out the approximate crisp value. Similarly, if you know the membership function and the crisp value, you can find out the membership degree.

Figure 4.1 illustrates a single fuzzy membership function for tall people; however, multiple such fuzzy membership functions on a domain can be presented in an integrated way, and each can be associated with a linguistic name. Figure 4.2 presents an integrated fuzzy membership function on a common domain considering the height of people.

In the case of the example shown in Figs. 4.1 and 4.2, 'Height' is the linguistic variable, and takes values such as 'Tall', 'Short' or 'Average'. A little variation on the definition introduces fuzzy hedges such as 'Not Very Tall' or 'Slightly Short'.

Membership functions (MFs) are the building blocks of fuzzy set theory and serve as key mechanism to map a crisp value into its equivalent fuzzy value and vice versa. Fuzzy membership functions may have different shapes, such as triangular, trapezoidal or Gaussian, depending on how the fuzzy membership is defined. Further, a membership function can be represented in either a continuous or discrete way. Let us see some types of the fuzzy membership functions.

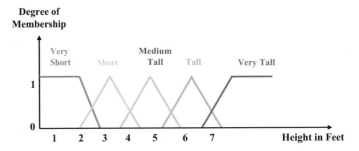

Fig. 4.2 Integrated fuzzy membership function

4.2.1   *Triangular Membership Function*

If the membership curve for a variable follows a triangular shape, the membership function is known as triangular membership function. Fuzzy function A is called triangular fuzzy function (A = (a, α, β)) with peak (or centre) a, left width α > 0 and right width β > 0 if its membership function has the following form:

$$A(x) = 1 - (a - x)/\alpha \text{ is if } a - \alpha \le x \le a$$
$$= 1 - (x - a)/\beta \quad \text{ if } a \le x \le a + \beta$$
$$= 0 \text{ otherwise}$$

See Fig. 4.3.

4.2.2   *Trapezoidal Membership Function*

Here, the membership function takes the form of a trapezoidal curve, which is a function of (A = (a, b, α, β).) with tolerance interval [a, b], left width α and right width β if its membership function has the following form:

$$A(x) = 1 - (a - x)/\alpha \text{ if } a - \alpha \le x \le a$$
$$= 1 \qquad\qquad \text{ if } a \le x \le b$$
$$= 1 - (x - b)/\beta \text{ if } a \le x \le b + \beta$$
$$= 0 \qquad\qquad \text{ otherwise}$$

See Fig. 4.4.

Fig. 4.3 Triangular fuzzy membership function

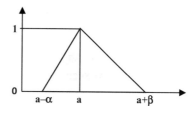

Fig. 4.4 Trapezoidal fuzzy membership function

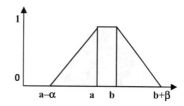

Fig. 4.5 Gaussian
membership function

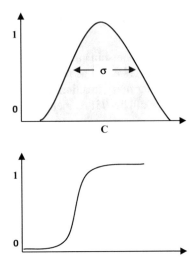

Fig. 4.6 Sigmoidal
membership function

4.2.3 Gaussian Membership Function

A Gaussian MF is determined by c and σ; where c represents the centre of the
membership function and σ determines width of the membership function. It is
defined as follows.

$A(x) = e^{(-(x-c)^2/2\sigma^2)}$; where c is the centre of the MF and σ is width of the MF.
See Fig. 4.5.

4.2.4 Sigmoidal Membership Function

A sigmoid function is defined as follows.

$A(x) = 1/(1 + \exp(-a(x-c)))'$; where a controls the slop at the crossover point
$x = c$.

A sigmoid function is open at one end, and hence it is very suitable to show
extreme situations such as 'very large'. See Fig. 4.6.

Real World Case 1: Fuzzy Decision Making in Tourism Sector
In the tourism sector, e-commerce is playing an important role to develop the
industry and improve services. On the other hand, combining e-commerce
technology with intelligent techniques has provided unique capabilities for
flexibility in fulfilling human needs. Tourism combined with fuzzy logic
approach and ecommerce technology has created further expansion of this

(continued)

industry, especially in better addressing customers' needs and tastes. The ultimate goal is to find a simple and applicable way to find a suitable tour plan, to suggest various available packages and to suggest an appropriate accommodation, by inputting data related to their interests and needs,. For instance, one could use a standard method for fuzzy decision-making and then try the Euclidean distance method, which is very simple in calculation.

4.3 Methods of Membership Value Assignment

Examples of fuzzy membership functions demonstrated in Figs. 4.1 and 4.2 are determined by the experience and observation of a persons' height during routine transactions. There are several methods to determine such membership functions. They can be based on common observations (as mentioned above), intuition, natural evolution, some logic (inference) and some rank.

The intuition method generally follows basic common sense knowledge. Based on one's personal experience and expertise, membership functions can be intuitively designed. Membership functions demonstrated in Figs. 4.1 and 4.2 are classic examples of this category.

Logic-based methods, such as inference and deductive reasoning, can also be used to develop membership functions. Genetic algorithm and neural networks are also used to obtain fuzzy membership functions. Genetic algorithm is used to evolve suitable membership functions from the randomly selected membership functions; and artificial neural network is used to cluster the data points.

Real World Case 2: Energy Consumption
Analytics can also be used for internal operations. Energy consumption accounts for roughly 65 % of the utility costs of a typical industry sector. However, costs can be well regulated by using energy more efficiently. At present times, smart data can help managers to build energy profiles for their enterprises. There are software solutions that gather data from multiple sources, including weather data, electricity rates and a building's energy consumption, to create a comprehensive building energy profile. Through a fuzzy-logic–based, predictive analytics algorithm, the software can fine-tune whether power comes from the grid or an onsite battery module.

4.4   Fuzzification and Defuzzification Methods

The process of converting a crisp value into its equivalent linguistic value through a well-defined membership function is known as fuzzification. In the example shown in Fig. 4.1, a crisp value of height x = 4 ft 3 in is to be converted to 'Tall people with the membership value 0.5. Similarly, the process of converting a linguistic value into its equivalent crisp value through a well-defined membership function is known as defuzzification. Table 4.1 summarizes the popular defuzzification methods.

4.5   Fuzzy Set Operations

Fuzzy sets also support operations such as union, intersection and complementation between the sets. Let A and B be fuzzy sets on the universe of discourse U. The operations of the fuzzy sets A and B are described as follows.

4.5.1   Union of Fuzzy Sets

Union of the fuzzy set = set A and B is denoted by A U B as defined below.

$$\mu_{AUB}(X) = \text{Max}\left[\mu_A(X), \mu_B(X)\right] \text{ for all } X \in U$$

See Fig. 4.7a showing the union between two fuzzy sets.

Table 4.1 Defuzzification methods

Method	Description
Centroid	Centroid method of defuzzification considers the curve represented by the fuzzy membership function and returns to the centre of area under the curve. This is the most appealing method for defuzzification. This method considers the composite output fuzzy set by taking the union of all clipped or scaled output. Fuzzy sets then derive the centre of mass of the shape represented by the composite output fuzzy set.
Centre of sums	Instead of building the output composite set using the union operation, the centre of sums method takes the SUM of the clipped/scaled output fuzzy sets and then computes the centre of mass of the resulting shape.
Singleton method	This simple method considers a crisp value associated with output membership functions, which are singleton, and calculates the weighted average of all qualified output sets.
Weighted average method	This method considers individual clipped or scaled output fuzzy sets, takes the peak value of each clipped/scaled output fuzzy set and builds the weighted (with respect to the peak heights) sum of these peak values.
Mean max method	This method finds the mean of the crisp values that correspond to the maximum fuzzy values.

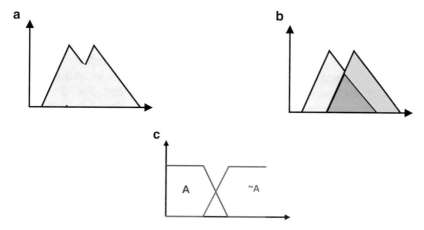

Fig. 4.7 Operations on fuzzy sets (**a**) Union of fuzzy sets, (**b**) Intersection of fuzzy sets, (**c**) Complement of a fuzzy set

4.5.2 Intersection of Fuzzy Sets

Intersection of fuzzy sets A and B is denoted by $A \wedge B$ as defined below.

$$\mu_{A \wedge B}(X) = \text{Min}\,[\mu_A(X),\ \mu_B(X)] \quad \text{for all } X \in U$$

Figure 4.7b illustrates the intersection between two fuzzy sets.

Optionally, instead of minimum operation, multiplication operation is also used to implement the intersection operation between two sets. In this case, the intersection operation is defined as follows.

$$\mu_{A \wedge B}(X) = \text{Mul}\,[\mu_A(X),\ \mu_B(X)] \quad \text{for all } X \in U$$

Further, if A and B are valid fuzzy sets, and if $A \subset B$ and $B \subset A$, then the sets A and B are said to be equal, denoted as $A = B$. That is, $A = B$ if and only if $\mu_A(x) = \mu_B(x)$ for $\forall\, x \in X$.

Beside minimum and multiplication, the frequently used fuzzy intersection definitions include probabilistic product, bounded product, drastic product and Hamacher's product.

4.5.3 Complement of a Fuzzy Set

Unlike the above two set operations, union of fuzzy sets and intersection of fuzzy sets, the complement operation does not require two operands, but a single operand. The complement of a fuzzy set is defined as follows.

$$\mu_{A'}(X) = 1 - \mu_A(X) \text{ for all } X \in U$$

The complement operation on a fuzzy set A is shown in Fig. 4.7c.

Operations such as algebraic sum, algebraic product, bounded difference and bounded sum operations on the fuzzy sets can also be defined by taking inspiration from the definitions of the equivalent classical operations. See Table 4.2 for definition of these operations.

See the following example illustrating the use of fuzzy set intersections.

Problem Statement   Consider the set of books B. The graded membership A of a book to determine affordability of a given book is defined as follows:

$$A = 0, \text{ if price of a book is more than 4999 \$}$$
$$A = 1 - \{price/500\} \text{ otherwise.}$$

Consider book prices in $ as 5000, 400, 300, 150, 400 and 100, respectively, for six books. Also consider a set of fuzzy quality indicators of the books (denoted as Q) as 1, 0.5, 0.8, 0.6, 0.2 and 0.3, respectively, for all the books in the set B. Calculate the

Table 4.2 Operations on fuzzy sets

Fuzzy Set Operation	Description	Definition
Union	Union of two fuzzy sets A and B	$\mu_{A \cup B}(X) = \text{Max}\,[\mu_A(X), \mu_B(X)]$ for all $X \in U$
Intersection	Interaction of two fuzzy set A and B	$\mu_{A \wedge B}(X) = \text{Min}\,[\mu_A(X), \mu_B(X)]$ for all $X \in U$
		Alternatively, operations like multiplication, probability products, bounded products and drastic products are used to implement fuzzy intersection.
Complement	Complement or negation of a fuzzy set A	$\mu_{A'}(X) = 1 - \mu_A(X)$ for all $X \in U$
Equality	Checks if the given fuzzy sets A and B are equal or not	If and only if $\mu_A(x) = \mu_B(x)$ for $\forall\, x \in X$.
Algebraic sum	Multiplication of two corresponding fuzzy values from fuzzy sets A and B is subtracted from their summation	Algebraic sum denoted as $\mu_{(A+B)}(X) = \mu_A(x) + \mu_B(X) - (\mu_A(x) * \mu_B(X))$] for all $X \in$ to U
Algebraic product	Multiplication of two corresponding fuzzy values from fuzzy sets A and B	Algebraic product denoted as $\mu_{(AB)}(X) = \mu_A(x) * \mu_B(X)$ for all $X \in U$
Bounded sum	Sum of two corresponding fuzzy values from fuzzy sets A and B; if exceeds 1 then restricted to 1	Bounded sum denoted as $\mu_{(A\,XOR\,B)}(X) = \text{Min}[1, \mu_A(x) + \mu_B(X)]$ for all $X \in U$
Bounded difference	Difference of two corresponding fuzzy values from fuzzy sets A and B; if goes beyond 0 then restricted to 0	Bounded difference denoted as $\mu_{(A\,O\,B)}(X) = \text{Max}\,[0, \mu_A(x) - \mu_B(X)]$ for all $X \in U$

Table 4.3 Book quality and affordability intersection

Book	Price ($)	Quality	Affordability	Worth to buy
Book 1	5000	1	0	0
Book 2	400	0.5	0.2	0.2
Book 3	300	0.6	0.4	0.4
Book 4	150	0.6	0.7	0.6
Book 5	400	0.2	0.2	0.2
Book 6	100	0.3	0.8	0.3

affordability of each book of the set of books B. Also find a book that is both most affordable and of high quality.

Solution The above problem deals with six different books. Let us name these books as Book1, Book2, Book3, Book4, Book5 and Book6. The fuzzy set showing graded membership towards 'Affordability' of these books is given as follows.

$$A = \{0/Book1, \ 0.2/Book2, \ 0.4/Book3, \ 0.7/Book4, \ 0.2/Book5, \ 0.8/Book6\}$$

The quality of the book denoted by Q is also give as follows.

$$Q = \{1/Book1, \ 0.5/Book2, \ 0.8/Book3, \ 0.6/Book4, \ 0.2/Book5, \ 0.3/Book6\}$$

The intersection of the fuzzy set A and fuzzy set Q is defined by minimum operations and intersection values are calculated (and placed under the column 'worth to buy') as shown in Table 4.3.

It is to be noted that the operations on fuzzy sets are well defined and crisp in nature. However, not only fuzzy sets, but also the operations on fuzzy sets can be fuzzy. Such fuzzy operators will be of great help while searching a large-volume database such as the Web. While searching, we typically use the Boolean operators like AND and OR, e.g., Black and White. In reality, while doing so, we mean more black and less white! The typical AND operation is very rigid and gives equal weight to both the colours. The OR is too soft, freely allowing both colours. The AND operation is to be diluted and the OR operation is to be made a bit strict. Such behaviour of the Boolean operators can be achieved by fuzzy AND and fuzzy OR operations. Such fuzzy operations may also be helpful while working with intelligent search engines on the Web.

4.6   Fuzzy Set Properties

Fuzzy sets follow properties such as commutativity, associativity, distributivity, idempotency, involution, transitivity, and Demorgan's laws. Fuzzy set properties are similar to crisp set properties, with a few exceptions. See Table 4.4 for a list of typical fuzzy set properties. Many such properties have been described by Takashi Mitsuishi et al. (2001).

Table 4.4 Properties of fuzzy sets

Property	Definition (where A and B are two valid fuzzy sets)
Commutativity	A U B = B U A and
	A ∧ B = B ∧ A
Associativity	A U (B U C) = (A U B) U C and
	A ∧ (B ∧ C) = (A ∧ B) ∧ C
Distributivity	A U (B ∧ C) = (A U B) ∧ (A U C) and
	A ∧ (B U C) = (A ∧ B) U (A ∧ C)
Idempotency	A U A = A and
	A ∧ A = A
Identity	A U Φ = A and A U {Universal Set } = {Universal Set} and
	A ∧ Φ = Φ and A ∧ {Universal Set } = A
Involution (Double Negation)	$(A')' = A$
Transitivity	If A < = B < = C then A < =C
Demorgan's Laws	$(A U B)' = A' ∧ B'$ and
	$(A ∧ B)' = A' U B'$

4.7 Fuzzy Relations

The relationship between two entities is generally crisp, stating whether the elements are related to each other or not. The strength of the relationship, however, is not generally measured. Examples of such crisp relationships are given below.

- Machine associated to an employee.
- Person married to.
- Currency with the country, e.g., USA with Dollar, India with Rupee.

All these examples illustrate the crisp relationship between two elements of different domains. If a person X is married to Y, until legal divorce, the person X is completely married to Y. In other words, either X is married to Y or not. There is no intermediate status.

Fuzzy relation identifies vague relation between entities of two different universes. A set of such vague relationship indicated in terms of the Cartesian product is also referred as a fuzzy set of relationships. The fuzzy relation is defined as follows.

A fuzzy relation R is a fuzzy set of the Cartesian product of classical sets $\{X_1, X_2, X_3, \ldots X_n\}$ where tuples $(x_1, x_2, x_3, \ldots, x_n)$ may have varying degrees of membership $\mu R(x_1, x_2, x_3, \ldots, x_n)$. That is,

$$R(X_1, X_2, X_3, \ldots X_n) = \int \mu R(x_1, x_2, x_3, \ldots, x_n) | (x_1, x_2, x_3, \ldots, x_n), x_i \text{ belongs to } X_i$$

Fig. 4.8 Fuzzy graph between X and Y sets

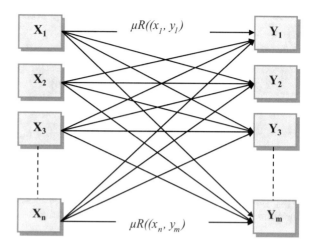

A fuzzy relation between two sets X and Y is also called a binary fuzzy relation and can be expressed as $R(X*Y)$ or $R(X,Y)$. If X and Y are equal, then it is denoted as $R(X^2)$. The fuzzy relation R on the sets X and Y, denoted by $R(X,Y)$ can be expressed in matrix form as follows.

$$R\,(X,\,Y) = \begin{cases} \mu_R\,(x_1,y_1), & \mu_R\,(x_1,y_2), & \ldots\ldots & \mu_R\,(x_1,y_m) \\ \mu_R\,(x_2,y_1), & \mu_R\,(x_2,y_2), & \ldots\ldots & \mu_R\,(x_2,y_m) \\ \ldots\ldots\ldots\ldots\ldots\ldots\ldots\ldots\ldots \\ \ldots\ldots\ldots\ldots\ldots\ldots\ldots \\ \mu_R\,(x_n,y_1), & \mu_R\,(x_n,y_2), & \ldots\ldots & \mu_R\,(x_n,y_m) \end{cases}$$

This matrix is also known as fuzzy relation matrix or simply fuzzy matrix for the fuzzy relation R. The fuzzy relation R takes values from [0, 1]. From such binary fuzzy relation, a fuzzy graph can also be developed. Each element of the first set X is connected with an element of another set Y. Such connection is shown graphically. Each connection has a weight factor, which is nothing but non-zero fuzzy values. See Fig. 4.8 for fuzzy graph structure.

On fuzzy relation, one can also apply operations such as union, intersection, and complement, along with other operations. The union is defined as maximum of the corresponding value, intersection is defined as minimum or multiplication of corresponding value and complement is defined as 1−the fuzzy value. Consider the following example.

4.7.1 Example of Operation on Fuzzy Relationship

Consider the set of machines, M, and the set of people, P, defined as follows:

M = { set of all machines in a domain}
e.g., M = $\{m_1, m_2, m_3, \ldots, m_n\}$ where n is a finite number; and
P = {set of people}
e.g., P = $\{p_1, p_2, p_3, \ldots, p_n\}$ where n is a finite number.

If machines of the set M are used by people of the set P, a relationship named R can be defined as a relationship of M*P and identified with the 'generally comfortable with' phrase. Here, R is a subset of M*P and denoted as R (subset of) M*P. The individual relationship can be presented as follows:

$$(p_1, m_1), (p_2, m_2), \ldots, (p_n, m_n)$$

Let us consider three persons, namely p_1, p_2, and p_3, and machines m_1, m_2, and m_3. Values of the 'generally comfortable with' (which is denoted by C), is also determined and presented in the matrix form, as shown in Table 4.5.

Let us consider another relationship between the same set of people and machines, stating the availability of required software and tools in the given machine. Let us identify this fuzzy relationship as A. The A relationship values are given in Table 4.6.

It is generally observed that people would like to work with the machine they are comfortable with. Simultaneously, they also need the necessary software for machine they are comfortable with. Here, the intersection operation can be used.

The intersection operation, denoted as (C ∧ A), is defined as minimum of the corresponding elements of C and A relationships, which is given below.

$$\mu C \wedge A\,(x, y) = \min(\mu C\,(x, y),\ \mu A\,(x, y))$$

The resulting matrix is given in Table 4.7.

The fuzzy relations possess properties such as projection, cylindrical extension, reflexivity, symmetry, transitivity, and similarity.

Table 4.5 'Generally comfortable with machine' relationship

C	m_1	m_2	m_3
p_1	1.0	0.4	0.7
p_2	0.3	1.0	0.6
p_3	0.7	0.6	1.0

Table 4.6 'Availability of the required software' relationship

A	m_1	m_2	m_3
p_1	0.8	0.3	1.0
p_2	1.0	0.0	0.4
p_3	0.4	0.8	0.1

Table 4.7 'Generally
comfortable' and
'Availability of software'
simultaneously

A	m_1	m_2	m_3
p_1	0.8	0.3	0.7
p_2	0.3	0.0	0.4
p_3	0.4	0.6	0.1

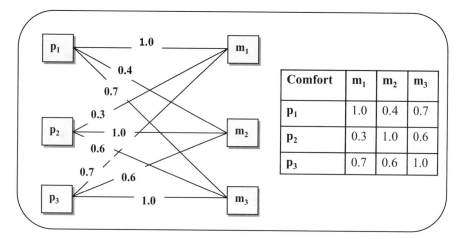

Comfort	m_1	m_2	m_3
p_1	1.0	0.4	0.7
p_2	0.3	1.0	0.6
p_3	0.7	0.6	1.0

Fig. 4.9 Fuzzy graph from the fuzzy relation 'Generally comfortable with machine'

For data science-related activities, fuzzy relations can be used to demonstrate relationship between the data and events using fuzzy graphs. Handling the large volume of information through visual graphs would make it easy to understand and communicate. A fuzzy graph is frequently expressed using a fuzzy matrix.

Mathematically, a graph G is defined as follows.

$G = (V, E)$; where V denotes the set of vertices and E denotes the set of edges.

If the values of the arcs of the graph G lies between [0,1], then the graph is fuzzy. If all values of arcs of the graph are crisp (either 0 or 1), then the graph is called a crisp graph. That is, a fuzzy graph is a function of vertices to the fuzzy values. It is defined as follows.

Fuzzy graph $G = (\sigma, \mu)$ is a pair of functions $\sigma : V \rightarrow [0,1]$ where V is the set of vertices and $\mu : V*V \rightarrow [0,1], \forall$ x, y $\in$ V.

Figure 4.9 illustrates the fuzzy graph of the fuzzy relation 'Generally Comfortable with Machine' given in Table 4.5.

4.8   Fuzzy Propositions

A statement is called proposition if it derives a truth value as positive (value 1) or negative (value 0). Following the same definition, if a statement derives a fuzzy truth value instead of crisp values such as 0 and 1, then the statement is called a fuzzy proposition. The truth value of the proposition P is be given as T(P). Instead of having only two crisp values, the T(P) has values within the range of [0,1]. Here is an example.

Let proposition P: Ms. X has good height. If Ms. X height is more than 5.4, then T(P) = 0.9. Another example can be Q: Ms. X is well experienced with T(Q) value 0.8. The T(P) and T(Q) values can be calculated as per the membership function associated with the fuzzy variable height and experience, respectively.

Many such fuzzy propositions are connected with each other logically to generate compound propositions. The following section introduces fuzzy connectives with examples.

4.8.1 Fuzzy Connectives

Negation ($\sim$), disjunctions ($\cap$), conjunction ($\cup$) and implications ($\Rightarrow$) are used as fuzzy connectives. The operations are defined as follows:

4.8.2 Disjunction

The disjunction function considers the maximum truth values from the given fuzzy proposition values. Hence, it is defined by the union function.

X is either A or B

Then X is A $\cup$ B

The meaning of 'X is A$\cup$B' is given by $\mu_A\cup_B$. The maximum operator as discussed earlier can be used for conjunction. An example is given. Consider p and q as follows:

$$P \rightarrow \text{Ms. X has good height.}$$
$$Q \rightarrow \text{Ms. X is well experienced.}$$
$$P \cup Q \rightarrow \text{Ms. X has either good height or efficient.}$$

Suppose Ms. X wants to be qualified in a sports category (such as basketball). She either needs to have good height or be well experienced (if she is not tall)

4.8.3 Conjunction

Conjunction generally refers to the simultaneous occurrence of events or an act of joining together. Hence, the meaning of conjunction is given by the intersection function as follows.

X is A,
X is B,
Then X is A ∩ B.

The meaning of 'X is A∩B' is given by $\mu_{A \cap B}$. The minimum operator as discussed earlier can be used for conjunction. Here is an example.

P → Ms X has good height.
Q → Ms X is well experienced.

Then the proposition produced by conjunction will be

P ∩ Q → Ms. X has good height and Ms. X is well experienced.

In this case, for Ms. X to be selected in a sports category, she must have the virtues of good height and experience.

4.8.4 Negation

The meaning of 'negation' is given be complement function, which is defined as: $1 - \mu A$. If we consider previous proposition p defined as 'Ms. X has good height'. The negation of p is given as follows.

P → Ms. X has good height

Then   ∼ P → Ms. X does not have good height.

4.8.5 Implication

The meaning of 'implication' is given by combination of negation and intersection function.

If X is A,
Y is B,
Then X is B.

The meaning of 'A$\Rightarrow$B' is given by max $(1-(\mu A)$ $(\mu B))$.

P $\rightarrow$ Ms. X has good height.
Q $\rightarrow$ Ms. X is well experienced.

Then P $\Rightarrow$ Q: if Ms. X is has good height then she is well experienced.

4.9   Fuzzy Inference

Fuzzy inference is a mechanism of mapping inputs with the respective outputs. It can be done in two ways: forward and backward. In the case of the forward mechanism, a set of available data is examined to verify the possibility to move towards a conclusion with the existing set of data. In the backward mechanism, for inference, a hypothesis is designed and matched against the available data. The fuzzy inference mechanism provides a powerful framework for reasoning with imprecise and uncertain information. The inference procedures are known as Generalized Modus Ponens (GMP) and Generalized Modus Tollens (GMT).

GMP is described as follows:

Rule: P is A, then Q is B
Given: P is A$'$
From the rule and given fact, one may conclude that Q is B$'$.

GMT is described as follows:

Rule: P is A, then Q is B
Given: Q is B$'$
One may conclude that Q is A$'$.

An example is given as follows:

The strawberry is very red.
If the strawberry is red then strawberry is ripe.
Then the conclusion can be 'the strawberry is very ripe'.

4.10   Fuzzy Rule-Based System

In case of the example shown in Figs. 4.1 and 4.2, 'Height' is the linguistic variable, which takes values such as 'Tall', 'Short, 'Average', 'Not Very Tall' or 'Slightly Short'. Such linguistic variables can be used in control structure such as if-then-else rules. Examples of some fuzzy rules are given as follows.

If person is 'Tall' then go to Game Zone A;
If height is 'Short' then go to Game Zone B.

The power and flexibility of simple if–then–else logic rules are enhanced by adding a linguistic parameter. Fuzzy rules are usually expressed in the form:

IF variable IS set THEN action.

The AND, OR, and NOT operators (also known as Zadeh's operators) defined in the earlier section are applicable to these rules. Such multiple rules are formed to define the proper logic to make intelligent and human-like decisions. Such logic based on fuzzy sets is called a fuzzy logic. A fuzzy rule can use more than one linguistic variable, each corresponding to a different fuzzy set. A system that uses the fuzzy rules as its main decision structure is known as a fuzzy rule-based system. A repository of such rules is known as the rule base of the system. In order to derive conclusions by the given rules of the system, an inference mechanism is used. The inference engine can use forward or backward chaining strategy to conclude and derive solutions from the available facts or hypothesis. The forward chaining applies the available data to the suitable rules from the rule base in order to reach the goal and completes the decision-making process. In backward chaining, a hypothesis is made and compared in reverse manner to match the existing data. Figure 4.10 illustrates the general structure, showing other components of a typical fuzzy logic-based system.

Fuzzy rule-based system models provided by Ebrahim Mamdani (1974) and Tomohiro Takagi and Michio Sugeno (1985) have also become popular to employ fuzzy logic for a rule-based system in a given business.

Further, instead of using the classical fuzzy logic representing a true 'many valued' logic about the partial truth, one can choose the 'limited multi-valued'

Fig. 4.10 Typical structure of a fuzzy rule-based system

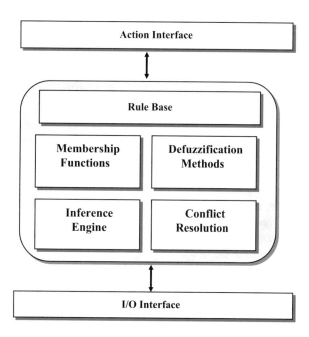

approach; e.g., seven-valued fuzzy logic, where values between 'true' and 'false' are divided into seven discrete categories. These categories are also provided generic descriptive names such as lowest, lower, low and medium, or seven domain-specific linguistic values. This can also be considered as a tiny contribution towards fuzzy logic.

Fuzzy logic basically maps the vague value into its equivalent crisp value through suitable membership function. That is, degree of youth (young person) is mapped to age and degree of tallness is mapped to height. However, while many attributes play a critical role in decision making, it is difficult to map them into their equivalent crisp values. For such variables, fuzzy to fuzzy mapping is required. This concept is identified as type 2 fuzzy logic. Type 2 fuzzy logic was originally experimented by Lotfi Zadeh (1975) to handle more uncertainty in a sophisticated way. It is obvious that a machine ultimately needs a crisp value, and hence a type reducer (or type converter) is required as an additional component (Sajja 2010).

> **Real World Case 3: A Personalized and Optimal Fitness Experience**
> Fuzzy Logic is very useful in situations to make decisions like a human. With fuzzy methodology, expert knowledge and other sports-related big data can be incorporated into wearable devices and then be persistently updated. Bringing together such ability along with data leveraged from the user's wearables and the user's known health conditions, data stored in the cloud will result in a personalized and optimal fitness user experience. For instance, a wearable is gathering real-time heart rate data and if it senses some type of abnormality with the user's heart rate, it should be able to alert the user to see a physician before a critical heart condition develops.

4.11   Fuzzy Logic for Data Science

Big or small, data are only useful if used in a smart way! Right skills, intelligent integration, the ability to see a bigger (and future) picture and goal driven processing can convert the data into smart data and hence a greater success. Many times data are incomplete and ambiguous. Fuzzy logic is helpful in completing the missing data and getting a better and broader picture of the situation. In isolation, the partial truths, or facts, mean nothing. A fuzzy logic approach helps in linking such data with the proper entities such as additional facts or inferred evidence. With such an approach, one may reach a nearly complete truth, which is a far more efficient process than the traditional ways. Further, the partial truth data are easier to reveal and they act as trail to the truth. This is the way that humans normally think. Data science activities such as data cleaning, processing, visualization and analysis can be benefited by the application of fuzzy logic.

Fuzzy logic has become popular because of its ability to handle linguistic parameters and its ability to deal with partial as well as uncertain information. By manipulating many uncertain data, one can judge the situation. In other words, the trails of partial truth lead us to the complete truth. Fuzzy logic can provide various benefits such as the following.

- It operates on linguistic variables, which is an advantage over traditional (pure symbolic or numeric) systems.
- It handles imprecise and partial data.
- It can complete the missing data.
- It has simplicity (to develop and to understand) and flexibility with a smaller number of rules to handle complex situations.
- It can model non-linear functions of arbitrary complexity.
- It has marketing advantages.
- It has a shorter development time.
- It is much closer to human thinking and reasoning.
- The nature of the techniques is robust; fuzzy techniques are not very sensitive to the changing environment.

Moreover, fuzzy logic enables us to efficiently and flexibly handle uncertainties in big data, thus enabling it to better satisfy the needs of real world big data applications and improve the quality of organizational databased decisions. Successful developments in this area have appeared in many different aspects, such as fuzzy data analysis technique and fuzzy data inference methods. Especially, the linguistic representation and processing power of fuzzy logic is a unique tool for elegantly bridging symbolic intelligence and numerical intelligence. Hence, fuzzy logic can help to extend/transfer learning in big data from the numerical data level to the knowledge rule level. Big data also contains a significant amount of unstructured, uncertain and imprecise data. For example, social media data is inherently uncertain.

This section highlights applications of fuzzy logic for managing various data on the Web by keeping the central objective of web mining in mind.

Real World Case 4: Stock Market Technical Analysis
Decision making practice in stock trading is complex. There are several technical indicators that are used by traders to study trends of the market and make buying and selling decisions based on their observations. This case is to deploy fuzzy inference to stock market, with four indicators used in technical analysis to aid in the decision making process, in order to deal with probability. The four technical indicators can be: Moving Average Convergence/Divergence, Relative Strength Index, Stochastic Oscillator and On-Balance Volume. The fuzzy rules are a combination of the trading rules for each of the indicators used as the input variables of the fuzzy system, and for all the four technical indicators used, the membership functions were also defined. The result is a recommendation to buy, sell or hold.

4.11.1   Application 1: Web Content Mining

Web content mining deals with mining of multi-media content by identifying and extracting concept hierarchies and relations from the Web in fully or partially automatic fashion. The fuzzy logic-based query can be designed, which is more natural and friendly for end users. Such fuzzy queries can use fuzzy linguistic variables such as near, average, about, and almost; along with domain variables. As stated earlier, Boolean operators for the queries in the search engine can also be diluted (in case of logical AND function) and strengthened (in case of logical OR function). Furthermore, the process of extracting information and mining useful patterns from the Web will become more interactive with the help of a fuzzy logic-based approach by accepting users' feedback in a fuzzy manner. Search results from the Web can be further filtered and mined for knowledge discovery. Dedicated techniques for text mining, image mining and general multi-media mining on the search outcome can also be thought of here. One exciting method for web mining can be described like this. In the first phase, the system accepts fuzzy and natural queries from its users. In the second phase, the system processes the query, generates a number of threads and assigns the threads to search the intended information on multiple search engines. In the third phase, the information from various search engines by various threads is collected at a common place and filtered / mined for the required knowledge. These phases are demonstrated in Fig. 4.11.

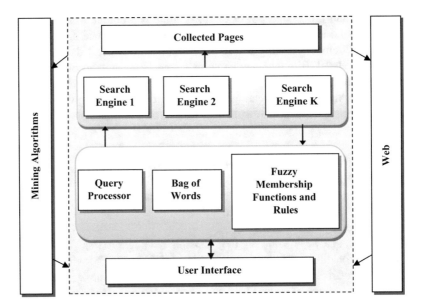

Fig. 4.11 Web content mining with fuzzy natural query

Not only the traditional Web, but also the semantic Web also can be benefited by the fuzzy logic approach. Tools like the fuzzy conceptual graph (FCG) may be used for content representation on the platform of the semantic Web, enabling the content to be ready for machine processing.

4.11.2   Application 2: Web Structure Mining

The objective of Web structure mining is to discover useful knowledge from the structure of the Web represented though hyperlinks. Hyperlinks exist in a web page and redirect the users to other pages/resources. One can form a tree or graph structure of such hyperlinks and traverse it in a meaningful way. The Web structure mining identifies relationship between the Web documents and presents the relationship in the form of such a graph of links. By accessing such graph or structure of the Web, one can calculate the relative importance of a webpage. The Google search engine uses a similar approach to find the relative importance of a webpage and orders the search result according to importance. This Web structure mining algorithm used by Google is known as the PageRank algorithm, and was invented by Google founders Sergey Brin and Lawrence Page (1998).

Fuzzy logic is generally used to classify large-sized data into clusters according to the usability of the content as per the user's interest as well as the nature of application. To facilitate such customized classification, a fuzzy user profile is generally used. With an interactive interface (or any dynamic approach), the system learns about the user's interests and models the information into a well-designed profile. If such data are on the Web platform, then data repositories generated by crawlers may be considered as an input source of data. A large crawler usually dumps the content of the pages it has fetched into a repository. It is not feasible in practice to dynamically fetch, compile and provide page ranks to the web pages. The page ranks should be pre-compiled and stored into the repository. This repository is used by many systems for classification of the content according to a topic or a theme, to construct a hyperlinked graph for ranking and analysis/mining of the content and so on. The proposed approach also applies to this repository. In general, webpage-related information such as metadata and robot.text files may contain fuzzy indicators showing the performance and usability of the content or the webpage. In this direction, if we think ahead, we can find different mechanisms to rank webpages using fuzzy logic.

For a modified page rank algorithm using fuzzy logic, the following can be done.

- Personalized add on filters can be set.
- Context dependency matrices may be developed.
- A mechanism of linguistic key words processing needs to be suggested to narrow down results.

Further, the page rank algorithm can be a weighted algorithm, where weights are personalized according to the user profile. Sometimes, a tag containing metadata is also assigned with the links. The edges of the graph can have fuzzy values as weights. These fuzzy values show the importance of the corresponding links. Since most of the databases and tools encode Web information in XML file format, fuzzy tags for the XML can be designed and mining techniques can be developed for such fuzzy XML documents. This is an impressive idea to mine the Web with help of fuzzy logic. Here, concentration is more on tags and structures, instead of the content. The same can be done for standard ontology-based documents. If XML or a standard ontology is not used to encode web information, one may think to mine the HTML documents on the Web. Fuzzy-directed graphs are also useful in web structure mining, returning an optimum path to surf the Web. With web structure mining, we can not only find out the previously unknown relationship between webpages, but also rank them according to relative importance using fuzzy logic. If the structure of a website (or a group of related websites) is known, it would be easy to navigate and recommend other useful sites and services. For example, structure mining can be useful to identify mapping between two business organizations sites, to thus help wholesalers and retailers as well as possible customers, and hence improve navigation and effective surfing of the Web. It can also help the business to attract more traffic and expand the business.

4.11.3   Application 3: Web Usage Mining

Web usage mining keeps track of the user's transactions and interactions on the web. Such a log of information is generally preserved at the server and presents information such as server access logs, user profiles, and queries of the information on the server. Web log mining can consider typical summaries such as request summary, domain summary, event summary, session summary, error and exception summary, referring organization summary and third party services or agents' summary. From the log, popular and useful site information is isolated and used for automatic suggestion and trend analysis. Here, instead of document clustering, site clustering and user clustering can be done in a fuzzy manner. Techniques such as web personalization (by considering the users log and usage), user profiles, session profiles, application profiles, trust-based recommender systems or intrusion detection can be further strengthened by a fuzzy logic-based approach. Figure 4.12 illustrates a sample application.

Fuzzy association rule mining techniques can also be used here for Web usage mining to determine which sets of URLs are normally requested together. This can be done in association with a corresponding fuzzy knowledge map or fuzzy cognitive map. Further, the fuzzy association rules may be used to predict the access path. Web usage mining can also be enriched with mining bookmarks, recommendations and email communications with fuzzy values.

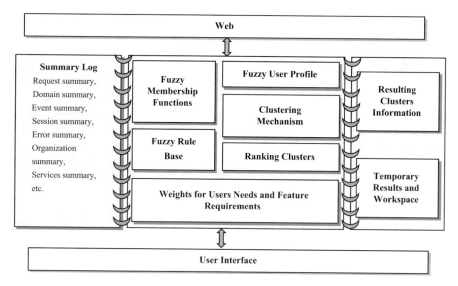

Fig. 4.12 Web personalization and clustering with fuzzy logic

4.11.4 Application 4: Environmental and Social Data Manipulation

Environmental data are difficult to handle because of their characteristics, such as their volume (spatial data and long-time series-based data) and heterogeneity (due to different data resources, different structures, objectives such as quantitative and qualitative measures, uncountable data such as number of fishes, and uncertainty through random variables and incomplete data). Such environmental data are incomparable due to different measures used in an approximate way or vague expert knowledge. To manage environmental data techniques such as fuzzy clustering and fuzzy kriging (Salski 1999), techniques are used in conjunction with the fuzzy knowledge-based modelling. In the knowledge base of the system, fuzzy rules that use fuzzy linguistic variables such as 'diversity of the vegetation type' and 'number of territories' with values like 'high', 'low' or 'average' are present.

Similarly, for remote sensing data, fuzzy logic can be used to analyse the remote sensing data for GIS ready information along with knowledge-based image interpretation. Data collected through a web sensor network can also undergo similar procedures. Fuzzy logic has been in use to detect heavy traffic using an intelligent traffic control system using environmental and infrastructural data (Krause, von Altrock and Pozybill 1997).

Fuzzy sets can also be used to mine opinions from social network platforms. Generally, social data are computationally and commonly represented by graph theory. Such a graph can have fuzzy weights for customized processing and traversing. It can also introduce a factor of trust while traversal and with recommendation of entity (specifically on internet of things [IoT]). This approach may also include the creation of fuzzy sets of sentiments between various connections in order to promote a product or entity.

For social network analysis, fuzzy metrics may be developed, as suggested by Naveen Ghali et al. (2012). Parameters such as closeness (degree with which an individual is nearer to all others in a network either directly or indirectly), density of the network (measure of the connectedness in a network), local or global centrality of a node in a network (how a node is tied with other nodes), and betweenness of the node in the network can be considered with corresponding fuzzy weights in the fuzzy metrics. Such metrics serve as very good tools for social network analysis. Similarly, performance of a social network can also be examined via fuzzy metrics having parameters such as effectiveness, efficiency and diversity.

Beside the above-mentioned scenarios, social network platforms have some inherent problems in general such as uncertainty and missing data, and thus require effective techniques for visualization. With fuzzy sets and graphs, it is possible to solve such problems to some extent. To identify terrorist activities on social network platforms, to see citations in authorship network such as DBLP,[1] and to recommend product based on fuzzy trust factors are example applications of fuzzy logic on a social network platform.

Another such area is that of medical data handling and mining. Medical data tend to be vague in nature and contain imprecision at a different level. Further, medical data include multi-media content, such as medical imaging and graphs.

Fuzzy logic can also be used in conjunction with other technologies, such as artificial neural network and genetic algorithms. Generally, artificial neural networks are used for classifying non-linearly separable data. In the case of complex and vague data, it is not advisable to directly provide such un-normalized data to the neural network. Here, fuzzy logic can help to convert the data into the desired form. A combination of artificial neural network and fuzzy logic provides the dual advantages of both fields. Further, the neural network does not have a capability of explanation and reasoning besides dealing with vague and uncertain data that can be efficiently handled by fuzzy logic.

4.12   Tools and Techniques for Doing Data Science with Fuzzy Logic

There are several tools available that aid in development of a fuzzy rule-based system. Java-based tools such as Funzy,[2] and fuzzy engine[3]; software tools based on R such as FRBS[4] and SFAD; python-based fuzzy tools such as pyfuzzy,[5] and

---

[1] dblp.uni-trier.de/

[2] https://code.google.com/p/funzy/

[3] http://fuzzyengine.sourceforge.net/

[4] http://cran.r-project.org/web/packages/frbs/index.html

[5] http://pyfuzzy.sourceforge.net/

other tools and libraries such as MatLab,[6] can be used to develop different fuzzy rule-based systems for a domain of interest.

Fuzzy Logic Tools (FLT) is another C++ framework that stores, analyses and designs generic multiple-input multiple-output fuzzy logic-based systems, based on the Takagi-Sugeno model (Takagi and Sugeno 1985).

BLIASoft Knowledge Discovery[7] is innovative software for data mining and decision-making that enables modelling, understanding and optimizing the most complex processes using fuzzy logic. It supports understanding, processing, modelling, optimizing and visualizing the data, and supports applications such as customer behaviour, fraud prediction, and forecasting of natural phenomena. This tool is basically a data mining tool; however, it provides support of fuzzy logic and artificial intelligence techniques for data science.

A machine learning framework is also available in order to develop understandable computational models from data.[8] It uses fuzzy logic-based machine learning methods through an object-oriented C++ programming language. It also provides the means to integrate with Mathematica software.

Software called Predictive Dynamix[9] provides computational intelligence tool for forecasting, predictive modelling, pattern recognition, classification, and optimization applications across various domains.

Real World Case 5: Emergency Management
Emergency management is one of the most challenging examples of decision making under conditions of uncertain, missing, and vague information. Even in extreme cases, where the nature of the calamity is known, preparedness plans are in place, and analysis, evaluation and simulations of the emergency management procedures have been performed, the amount and magnitude of shock that comes with the calamity pose an enormous demand. Hence, the key for improving emergency preparedness and mitigation capabilities is utilizing rigorous methods for data collection, information processing, and decision making under uncertainty. Fuzzy logic-based approaches are one of the most capable techniques for emergency mitigation. The advantage of the fuzzy logic-based approach is that it enables keeping account of incidents with perceived low possibility of occurrence via low fuzzy membership, and updating these values as new information is gathered.

---

[6]http://www.mathlab.mtu.edu/

[7]http://www.bliasoft.com/data-mining-software-examples.html

[8]http://www.unisoftwareplus.com/products/mlf/

[9]http://www.predx.com/

4.13 Exercises

1. State the features of membership functions.
2. Outline what is meant by the terms fuzzy set theory and fuzzy logic, and how they can be used for the treatment of uncertainty in artificial intelligence (AI) systems. To what extent can they be used for all types of uncertainty, or are they only useful for particular types?
3. Explain the difference between randomness and fuzziness.
4. Explain the method of generating membership function by means of neural networks and genetic algorithm.
5. Elaborate on the following sentence: 'Fuzzy logic fits best when variables are continuous and/or mathematical models do not exist'.
6. Consider the one-dimensional data set 1, 3, 4, 5, 8, 10, 11, 12. Let us process this data set with fuzzy (c-means) clustering using $c = 2$ (two clusters) and the fuzzifier $m = 2$. Assume that the cluster centres are initialized to 1 and 5. Execute one step of alternating optimization as it is used for fuzzy clustering, i.e., a) compute the membership degrees of the data points for the initial cluster centres; b) compute new cluster centres from the membership degrees that have been obtained before.
7. **(Project)** Investigate and justify various advantages and disadvantages of fuzzy logic in real-world problem solving.
8. **(Project)** Forecasting is one of the strengths of data mining and enables enterprises to better plan to exceed the needs of its business. Forecasting enables more efficient recruitment, purchasing, preparation and planning. Consider an enterprise/sector of your choice (e.g., hotel industry) and utilize a fuzzy logic approach to support the business forecasting.

References

Brin, S., & Page, L. (1998). The anatomy of a large-scale hypertextual web search engine. *Computer Networks and ISDN Systems, 30*(1–7), 107–117.

Ghali, N., Panda, M., Hassanien, A. E., Abraham, A., & Snasel, V. (2012). Social networks analysis: Tools, measures. In A. Abraham (Ed.), *Social networks analysis: Tools, measures* (pp. 3–23). London: Springer.

Krause, B., von Altrock, C., & Pozybill, M. (1997). Fuzzy logic data analysis of environmental data for traffic control. *6th IEEE international conference on fuzzy systems* (pp. 835–838). Aachen: IEEE.

Mamdani, E. H. (1974). Applications of fuzzy algorithm for control of simple dynamic plant. *The Proceedings of the Institution of Electrical Engineers, 121*(12), 1585–1588.

Mitsuishi, T., Wasaki, K., & Shidama, Y. (2001). Basic properties of fuzzy set operation and membership function. *Formalized Mathematics, 9*(2), 357–362.

Sajja, P. S. (2010). Type-2 fuzzy interface for artificial neural network. In K. Anbumani & R. Nedunchezhian (Eds.), *Soft computing applications for database technologies: Techniques and issues* (pp. 72–92). Hershey: IGI Global Book Publishing.

Salski, A. (1999). Fuzzy logic approach to data analysis and ecological modelling. *7th European congress on intelligent techniques & soft computing.* Aachen: Verlag Mainz.

Takagi, T., & Sugeno, M. (1985). Fuzzy identification of systems and its applications to modeling and control. *IEEE Transactions on Systems, Man, and Cybernetics, 15*(1), 116–132.

Zadeh, L. A. (1965). Fuzzy sets. *Information and Control, 8*, 338–353.

Zadeh, L. A. (1975). The concept of a linguistic variable and its application to approximate reasoning. *Information Sciences, 8*, 199–249.

Chapter 5
Artificial Neural Network

5.1 Introduction

Intelligence is a key resource in acting effectively on problems and audiences. Whatever the business is, if it is done with added intelligence and insight, it can provide high rewards and increases in quality of product, services and decisions. Intelligence can be defined as an ability to acquire knowledge as well as having wisdom to apply knowledge and skills in the right way. It is also defined as an ability to respond quickly, flexibly and by identifying similarities in dissimilar solutions and dissimilarity in similar situations. Some mundane actions such as balancing, language understanding and perception are considered as highly intelligent activities; these actions are difficult for machines. Some complex actions by animals, on other hand, are considered as non-intelligent activities. An interesting experiment has been carried out on the wasp (an insect that is neither bee nor ant, but similar to these two), which behaves in very complicated way while searching and preserving food. The experiment is described on a website presenting reference articles to Alan Turing.[1] According this source, a female wasp collects food, puts it near its burrow, and goes inside the burrow to check for intruders. If everything is safe, the wasp comes out and puts the food into the burrow. During the experiment, the food is moved a few inches from its original place. Instead of finding the food just a few inches away, the wasp goes in search of new food, again puts the food near the burrow and repeats the procedure. This behaviour is complex, however, non-intelligent. Besides the aforementioned mundane tasks, there are expert problem solving and scientific tasks such as theorem proving, fault finding and game playing which also come under the intelligent category.

Sometimes it is expected for a machine to behave in intelligent manner in the absence of human control and/or to relieve humans from the burden of decision

---

[1] http://www.alanturing.net/

© Springer International Publishing Switzerland 2016
R. Akerkar, P.S. Sajja, *Intelligent Techniques for Data Science*,
DOI 10.1007/978-3-319-29206-9_5

making. The field that deals with the efforts to make machines intelligent is known as artificial intelligence. As stated in previous chapters, artificial intelligence (AI) is a loosely defined field. It studies the abilities and characteristics that make human intelligent; identifies the tasks where human are better; and tries to simulate these activities through machines. Artificial intelligence uses heuristic methods (practical, rule of thumb methods), symbolic manipulation and non-algorithmic approaches.

It is obvious that intelligence is a key to managing complex and large business. To support business activities and to manage data related to the business in an intelligent as well as automated way, support of intelligent systems is needed. Though intelligent systems such as experts systems are very helpful, they come with a bundle of difficulties. Prime difficulties with intelligent systems are the abstract and dynamic nature of knowledge, the volume of knowledge, limitations of knowledge acquisition and representation, and lack of guidelines for development models as well as standards. There are bio-inspired techniques available that can learn from data, provided suitable structure and example data sets are available. Artificial neural network (ANN) is one such bio-inspired techniques. It supports self-learning from complex and voluminous data; hence, it is most suitable for handling big data of the business. Such a technique can be used to highlight hidden patterns in complex data, classifies the large volume of data into various classes (with the help of predefined class definitions or even without it) and highlights the meaningful patterns from the data.

This chapter illustrates difficulties with the symbolic learning techniques and introduces artificial neural network as bio-inspired techniques. The chapter discusses various models to design an artificial neural network, including the Hopfield model, perceptron model and its variations, and the Kohonen model. The chapter also discusses guidelines and practical heuristics, which will be very helpful in designing a neural network for a selected domain. The tools and utilities to implement the designed neural network are also given at the end. There is also a list including free and open source tools and packages. To demonstrate how the neural network technique can be used to manage a bundle of complex data on a social network platform, a case of emotions mining on the Web is discussed. A broad architecture and structure of a neural network and training data set examples are also provided for those of you who want to implement the emotions mining on social network platform. Artificial neural network is the technology that helps in automatically learning from a large volume of data just by providing suitable structure and sample training data. We have also discussed the problem of overfitting. This is a very active area of research.

5.2 Symbolic Learning Methods

It is obvious (as stated in earlier chapters) that the intelligence requires various types of knowledge. Some examples of knowledge are domain knowledge, common sense knowledge, knowledge about knowledge (meta-knowledge), tacit knowledge, and explicit knowledge. It is necessary to collect an ample amount of knowledge and store it in a repository. Typically, procedures used to collect knowledge are known as knowledge acquisition procedures. Due to the complex nature and size of knowledge, it is hard to fully automate a knowledge acquisition procedure. Once the knowledge is collected, it has to be stored into knowledge structure after some preliminary processing and validations. Knowledge structures such as rules, semantic net, frames and scripts may be utilized for the same. One may also choose hybrid knowledge representation structure such as embedding rules in frames and scripts. An inference mechanism (in order to refer to existing knowledge and to infer new knowledge) is designed. As knowledge is represented into the repository, called a knowledge base in a symbolic manner, such representation of knowledge is known as symbolic representation of knowledge. Furthermore, components such as explanation and reasoning, user interface and self-learning also need to be developed. After completing the development and integration procedure, a knowledge-based system is tested with suitable test cases. These phases are demonstrated in Fig. 5.1.

This is a long procedure requiring handsome effort. However, it is possible that the knowledge may quickly become obsolete and effort of the development of the system are in vain. Furthermore, there are many problems and limitations in developing such systems. The very first problem is the lack of proper knowledge acquisition methods and tools. We have very few knowledge acquisition methods. Mostly we have support from fact-finding methods such as interview, questionnaire, record reviews and observations. Similarly, we do not have efficient knowledge representation structures and inference mechanism. The lack of computer-aided software (CASE) tools also creates some limitations when it comes to the development of knowledge-based systems.

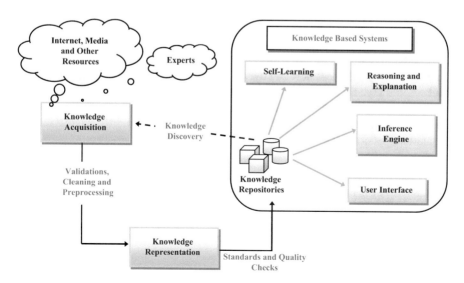

Fig. 5.1 Intelligent system development using symbolic representation of knowledge

Table 5.1 Major difficulties of symbolic representation of knowledge

Limitations	Description
Nature of knowledge	The nature of knowledge is very abstract and complex.
	It is dynamic in nature and continuously keeps changing.
	Knowledge is hard to characterize.
Volume of knowledge	To solve even a simple problem, a high volume of knowledge is required; e.g., to make a move on chess playboard, complete knowledge of the game is required.
Knowledge acquisition	A few knowledge acquisition methods are available.
	Due to the nature of knowledge and types (such as tacit and subconscious knowledge), it is difficult to the automate knowledge acquisition process.
	The main source of knowledge is domain experts, which are rare commodities of the field.
Knowledge representation	It is the knowledge engineer's knowledge that is reflected into the knowledge base.
	Tacit, implicit and subconscious knowledge are difficult to represent.
	It is required to represent the high volume of knowledge efficiently into a knowledge structure, which creates hurdles in the storage of knowledge.
Tools and guidelines	There is little support available in the form of overall development models, approaches and tools.
Dynamic data accommodation	Once the required knowledge is collected, it is difficult to change the knowledge base content. Through self-learning and inference mechanism, new knowledge is generated; however, external data are difficult to manage.
Cost-benefit ration	The collected knowledge soon becomes obsolete. In this case, the cost of development of such a system is high compared to its intended benefits.

In many situations, data are available, but it is very difficult to deduce generalized rules from the set of data. It can be concluded that the quality of such a system depends on the understanding and grasping of the knowledge engineer (or experts) who identifies the generic knowledge from the collected data. Furthermore, a knowledge-based system, once developed using symbolic knowledge representation, does not incorporate new and dynamic data. Above all, the development effort and cost are high for such a system. These difficulties are illustrated in Table 5.1.

To overcome the above limitations, an approach that learns automatically from the data when provided a broad architecture along with lots of sample data is suggested. Section 5.3 describes the approach.

5.3 Artificial Neural Network and Its Characteristics

As stated in the previous section, the symbolic representation of knowledge has some limitations. To overcome these limitations, an *artificial neural network*-based approach is used.

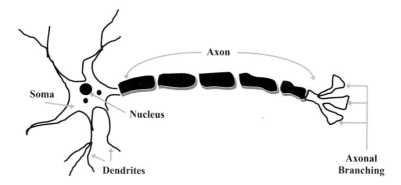

Fig. 5.2 Biological neuron

An *artificial neural network* is a predictive model inspired by the way the brain operates. Let us imagine that the human brain is a collection of neurons wired together. Each neuron looks at the outputs of the other neurons that feed into it, does a calculation, and then either fires (if the calculation exceeds some threshold) or does not fire. Artificial neural networks can solve a large variety of problems such as handwriting recognition and face detection, and they are used profoundly in deep learning, one of the subfields of data science.

An artificial neural network is made up of large number of neurons interconnected with each other. Each neuron is a simulation of a biological neuron of the human nervous system. A biological neuron is considered as a basic unit of the nervous system. A biological neuron is a cell having a nucleolus, cell body, and connections to other neurons of the nervous system. The structure of a typical biological neuron is shown in Fig. 5.2.

A biological neuron receives signals through its dendrites either from connected neurons or from other sensory inputs of the human body, as shown in Fig. 5.2. The cell body, which is also known as the soma, integrates (or processes) such multiple inputs from different dendrites in a spatial and temporal manner. When sufficient input is received and a threshold value is reached, the neuron generates an action (spike) and it fires some outputs to the connecting neurons. If there is no input or an insufficient amount of input, the collected inputs gradually vanish and no action is taken by the neuron. This function is simulated to build artificial neurons. An artificial neuron has a small processing mechanism (function) as its nucleus. It may receive n inputs through n connections. These connections are weighted connections. The core function provided as the nucleus to the artificial neuron sums up the inputs using the weights of connections as significance of the inputs. If the resulting sum is high enough, it may fire some output to the connecting neurons. The structure of an artificial neuron is illustrated in Fig. 5.3.

As illustrated in Fig. 5.3, an artificial neuron can also be considered as a device with many inputs and an output. A neuron works in two modes, use mode and train mode. A neuron must be trained with a sufficient number of samples before it can

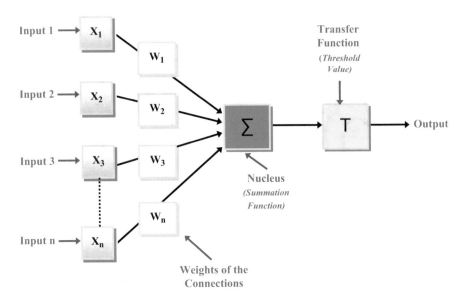

Fig. 5.3 Structure of an artificial neuron

be used. To train a neuron, along with a sample data set, the learning strategy also must be determined. That is, the neuron must know when to fire (send output) and when not to fire. The firing rule determines whether a neuron should fire for any input pattern.

Such a large number of neurons is interconnected through weighted connections between individual neurons. These neurons work in a very simple way, but all together—in a parallel fashion. The abilities of making intelligent decision making and learning come from their parallel working. Each neuron calculates a tiny thing at its local level, which at the end is summed up as a global solution. That is, processing and control are distributed in such a network. Furthermore, a large number of neurons is operating in a parallel and asynchronized manner in a neural network, and if some of the neurons are not working, the global solutions will not be much affected. This is the situation, where even though a few small pieces of a jigsaw puzzle are not found, the overall picture is clear in the viewer's mind! This characteristic of a neural network is known as fault tolerance. Since there are so many neurons in a neural network, the network can afford to miss some neurons. Table 5.2 lists the major characteristics of an artificial neural network.

An artificial neural network has the ability to learn from data provided to it. Typically, such networks perform by forming systematic structures and layers. Historically, McCulloch and Pitts (1943) developed the first neural model. After that, Frank Rosenblatt (1957) studied a model of eye and brain neurons and developed a model. Since eye and brain neurons are mainly involved in perception-

Table 5.2 Major characteristics of an artificial neural network

Characteristics	Description
Large number of neurons	An artificial neural network encompasses a large number of processing units called neurons. Each neuron contributes towards a global solution using the function stored within its core part (nucleus), inputs provided to it, and weights associated with the connections.
Weighted connections	Every connection, from neuron to neuron and from input to neuron, is associated with a value showing the strength of the connection.
Parallel working	All the neurons, according to the values, weights and core processing functions provided within the neurons, work in parallel.
Asynchronous control	Since all the neurons works in parallel manner, the processing as well as control is asynchronous in nature. With the help of such a working mechanism, each neuron can work in an independent manner, and still be able to contribute towards the global solution.
Fault tolerance	Since many neurons work together, and contribute little in a simple way, if some are not working, the network still operates.

related activities, the model was named Perceptron. Later, Marvin Minsky and Seymour Papert (1969) provided mathematical proofs of the limitations of the perceptron by presenting the classic exclusive-OR (XOR) problem and the inability of the perceptron model to solve it. Later, the model of the multi-layer Perceptron was proposed by the Rosenblatt. Besides this model, there are other models and paradigms, such as Boltzmann machine (Ackley et al. 1985), Hopfield's network (Hopfield 1982) Kohonen's network (Kohonen 1988), Rumelhart's competitive learning model (Rumelhart and Zipser 1985), Fukushima's model (Fukushima 1988), and Carpenter and Grossberg's Adaptive Resonance Theory model (Carpenter and Grossberg 1988).

All these models encompass a large number of neurons with weighted connections in different arrangements and learning mechanisms. The learning mechanism of a neural network is concerned with the behaviour of the neuron and its strength (weight) of connections to produce desirable output. The choice of a neural network model and appropriate learning algorithm is a major issue in employing an artificial neural network. Some of the above-mentioned models are described in next section.

5.4 ANN Models

This section mainly describes the Hopfield model, a perceptron by Rosenblatt, a multi-layer perceptron model along with other major models of ANN with the learning paradigms.

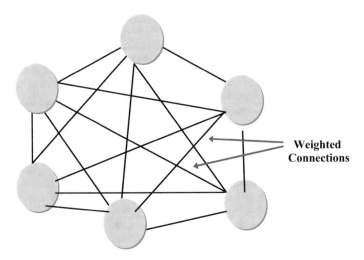

Fig. 5.4 An example of Hopfield network

5.4.1   Hopfield Model

The Hopfield model of artificial neural network was proposed by John Hopfield in 1982. A Hopfield neural network model consists of many neurons connected with each other in such a way that each neuron is connected with every other neuron. For each connection, a weight factor is associated, called connection weight. The connection weight from neuron i to neuron j is given by a number w_{ij}. The connection weights for all the connections between the neurons are represented in a matrix format. This matrix is called W. The collection of all such numbers is represented by the weight matrix W. The weights in the matrix W are denoted as w_{ij}. These weights in matrix W are symmetric in nature. That is, w_{ij} is same as w_{ij} for all i and j. Here, each node plays the role of input unit as well as output unit. Figure 5.4 represents a small Hopfield neural network.

The Hopfield neural network learns in the following way:

- Each neuron of the Hopfield network is considered as either 'Active' or 'Inactive'.
- A weighted connections matrix is established between the neurons in such a way that each node is connected with every other node of the network.
- A random node is chosen.
- If any of the neurons neighbours are active, the sum of the weights of the connections to the active neighbours are computed.
- If the sum is positive, the unit becomes active or else inactive.
- The process is repeated until the network reaches a stable state.

Learning in the Hopfield network can be done in an asynchronous or synchronous manner. If a neuron is picked and the status of the neuron is updated immediately by calculating the sum of the weights from active neighbours; then it is known as asynchronous learning. In the case of synchronous learning, the sum is calculated for all the neurons simultaneously. The Hopfield network is used for modelling associative memory and pattern matching experiments.

5.4.2 Perceptron Model

As stated earlier, eye and brain neurons were considered to model neural network architecture by a scientist called Frank Rosenblatt. This model is known as Perceptron. The simplest perceptron can be build using a single neuron. Each neuron has its processing function as core component. Besides this core component, there are many inputs of variable strength. The Perceptron receives input from all its connections; processes the inputs by considering weight (strength) of the connection from which an input is coming; and identifies significance of the output generated. If the output is significant and exceeds from a given threshold value, then the perceptron fires the calculated output. The core function that processes the inputs is known as an activation function. Figure 5.5 illustrates an example model of a neural network.

The above example discusses the classic problem of to be or not to be! Consider a case where a son wants to join the army and his parents have different opinions about the decision. The son is closely and strongly related to his mother. In this example, the strength of the mother–son relationship is 0.8. The connection is named

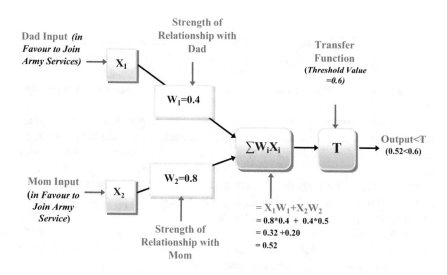

Fig. 5.5 An example of perceptron: to be or not to be

as X_1; and the weight is denoted as W_1. The son's relationship with the father is comparatively weak. Here it is 0.4. The connection is named as X_2; and the weight is denoted as W_2. The activation function used by the Perceptron shown in Fig. 5.5 is given below.

$$\text{Activation function} = \sum W_i X_i \text{ for } i = 1, 2.$$

When mother strongly believes that the son should not join army, then the input from mother is, say, 0.4 (X_1). Similarly, the value from the father is 0.5(X_2). The activation function processes these inputs as follows.

$$\begin{aligned}
\text{Activation function} &= \sum W_i X_i \text{ for } i = 1, 2. \\
&= W_1 X_1 + W_2 X_2 \\
&= 0.8 * 0.4 + 0.4 * 0.5 \\
&= 0.32 + 0.20 \\
&< 6.0 \text{ (Threshold value)}
\end{aligned}$$

The activation function in the illustrated case calculates a value of 0.52, which is less than the threshold value set for the perceptron. That means the perceptron does not fire. The output may be considered as null or zero. That is, from Fig. 5.5, it is obvious that the son's relationship with his mother is stronger; hence, she can influence her son's decisions.

To determine when a perceptron should be fired, a formal function can be defined as follows.

$$Y = \begin{cases} +1 \text{ if the output is greater than zero (or a threshold value)}. \\ -1 \text{ if the output is less or equal to zero (or a threshold value)}. \end{cases}$$

Here, y is an action taken by the perceptron. Y represents either that the perceptron fires or that it does not.

A similar perceptron can be designed for two inputs, logical AND gate. A perceptron used for simulation of logical AND gate actually identifies two regions separated by a line. See Fig. 5.6a. This part of Fig. 5.6 illustrates the behaviour of a perceptron graphically. It shows that how the perceptron has identified the line separating the two categories of solutions: output 1 is represented by filled circle and output 0 is represented by hollow circle. Figure 5.6b presents calculations of the function, considering w1 and w2 as 0.5.

Just by changing the weights and/or threshold, the same perceptron can be used to simulate logical OR function. Figure 5.6c and d represent the changed perceptron and the behaviour of the perceptron, respectively.

From the above example, it can be generalized that a single perceptron can be used to solve a linearly separable problem. Figure 5.7a graphically illustrates a

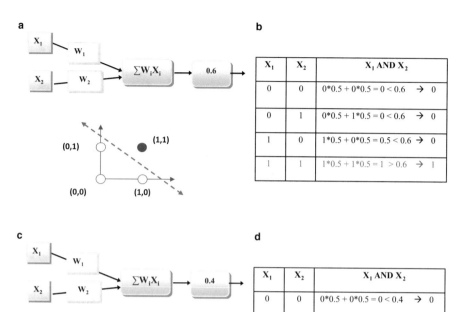

Fig. 5.6 Perceptron for Boolean functions AND and OR

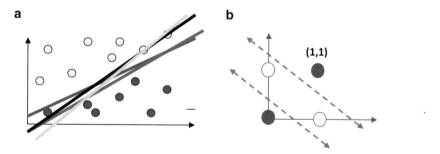

Fig. 5.7 Linearly separable problem and XOR

category of linearly separable problems. However, according to Marvin Minsky and Seymour Papert (1969), it is impossible for a single perceptron network to learn non-linear functions such as exclusive OR (XOR); see Fig. 5.7b. To solve such problems, a multi-layer perceptron or different combination of neurons may be used.

Real World Case 1: Energy Consumption Using Artificial Neural Network
It is important to forecast a typical household daily consumption in order to design and size suitable renewable energy systems and energy storage. ANNs are recognized to be a potential methodology for modelling hourly and daily energy consumption and load forecasting.

5.4.3 Multi-Layer Perceptron

This type of neural networks consists of multiple layers of neurons. Typically, there are three categories of layers: (1) an input layer, (2) one or more hidden layers and (3) an output layer.

All the neurons in the network are connected with each other in such a way that every neuron is connected with other neurons in the adjacent layer in a forward fashion. In a way, the network is considered to be fully connected, as each neuron of a given layer is connected with every neuron. All the input layer neurons are connected to the immediate hidden layer neurons.

Input neurons have values from the input interface. All the hidden neurons contain a processing function called the hidden function, which is a function of weights from the previous layer and previous values. A hidden function is any suitable activation function such as a sigmoid function. The hidden neurons outputs are forwarded to the output layers. The output layer neurons also contain an output function in each neuron. The output function is a function of weights from hidden layer neurons and values provided by the hidden nodes. In many applications, the internal neurons of the network apply a sigmoid function as an activation function. Figure 5.8 represents the general structure of a multi-layer perceptron.

The multi-layer perceptron is designed, connected and trained in a systematic manner, which is described in next section.

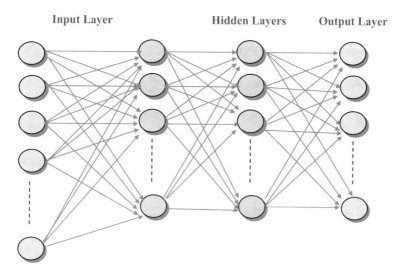

Fig. 5.8 General structure of the multi-layer perceptron

Real World Case 2: The Use of Artificial Neural Networks in On-Line Booking for the Hotel Industry

Three layers of multi-layer perceptron (MLP) ANN are adopted, and are trained using information from previous customers' reservations. Performances of ANNs should be analysed. They act in a rather reasonable way in managing the booking service in a hotel. The customer requires single or double rooms, while the system gives as a reply the confirmation of the required services, if available. Moreover, in the case that rooms or services are not at disposal, the system proposes alternative accommodations using an ANN approach.

5.4.3.1 Designing the Neural Network Phase

The very first phase of a neural network is to determine the number of input, output and hidden nodes. Typically, there is one input layer and one output layer. Selection of the number of hidden layers is mostly based on trial and error; however, there is a popular heuristic that one always starts with one or two hidden layers. The number of neurons in the hidden layers is somehow between the input layer neurons and output layer neurons.

The input layer of a neural network consists of a number of neurons according to the input opportunities. Critical parameters that play a crucial role in determining output must be identified. Let these parameters be the set of X, where X is defined as follows.

$$X = \{x1, \ x2, x3, \ ...., xn\}.$$

Design an input layer with the above listed X with n neurons.

Similarly, determine the output opportunities. Let the set of all opportunities be O; where O is defined as follows.

$$O = \{o1, \ o2, o3, \ ...., om\}.$$

Design an output layer containing m neurons as denoted in the set O. Each output node has an output function. The output function is a function of weights towards the output layer neurons with the hidden values calculated.

Also consider two hidden layers. The number of neurons in the hidden layer may be finalized with a typical heuristic as follows.

$$\text{Number of hidden neurons} = (m + n)\,/2;$$

where m is the total number of neurons in the output layer; and n is the total number of neurons in the input layer.

One may consider a different heuristic such as:

$$\text{Number of hidden neurons} = (m + n) * (2/3);$$

where m is the total number of neurons in the output layer; and n is the total number of neurons in the input layer.

Each hidden node has a hidden function within it. The hidden function is a function of weights to the hidden nodes with the input values.

5.4.3.2   Connecting Neurons in Feed-Forward Manner

After determining the design of the input layer, hidden layers and output layer with respective processing functions, the neurons need to be connected with each other. Connect all the neurons of the network in such a way that each neuron is connected with every neuron of the immediate layer in the forward direction only.

After connecting neurons, assign random weights to these connections. After connections, the neural network is ready to learn.

5.4.3.3   Learning Phase

A neural network learns by processing training data. The training data provided to the neural network plays an important role and determines the quality of the knowledge stored in terms of weights in its connections. The training data should have many sets of input as well as the corresponding output.

The neural network learns in the following way.

- Consider the first set of data. Give only input values to the input layers.
- Let your network compute what it thinks.
- Compare the calculated output with the correct output given in the training data set.
- Find errors using the typical back propagation learning algorithm and adjust the weights.
- Repeat this until the network gives correct output for the data set or the error is acceptably small.
- Repeat the procedure for all the data sets.

This is a broad outline of supervised learning using a back propagation learning paradigm. Since the network is learning from the data, the quality of the neural network is directly dependent on the data. If data sets are poor and considering insignificant cases, the network learns to make decisions as demonstrated in the data. Furthermore, if data are good, but do not cover all categories of decision-making examples, the neural network would be biased towards only the given types of data. When new types of data are experienced in reality, the network may not be able to provide the correct output.

There are many other ways through which a neural network can learn. Other popular paradigms include unsupervised learning and reinforcement learning. In the case of these learning approaches, a valid and correct data set to train network is not completely or partially available. For example, in the case of reinforcement learning, the network is not provided correct output, but rather rewards, if the learning is going in the desired direction. The reward can be a hint towards the next action. In unsupervised learning, the data have no target attribute. Historically, clustering has been the obvious technique for unsupervised learning.

5.4.4   Deep Learning in Multi-Layer Perceptron

As explained in a previous section, generally there are one or two hidden layers in the multi-layer perceptron. It is clear that if there is no hidden layer, a perceptron can solve only linearly separable problems. See the example given in Figs. 5.6 and 5.7a. If there is one hidden layer, the ANN can approximate a function that has a continuous mapping from one finite space to another. If there are two hidden layers, the neural network can approximate any smooth mapping to any accuracy. That is, introduction of the hidden layer helps the network to exhibit non-linear behaviour. Though the hidden layers do not interact directly with the input or output layer, they can influence the working of ANN. The idea of adding many hidden layers in a multi-layer perceptron seems promising to achieve complex and non-linear behaviour; but the additional hidden layers cause problems in learning. That is, adding more hidden layers can help in learning in a perfect non-linear manner, but it also causes overfitting the network and increases the issues related to efficiency. Here, the earlier layers can learn better, but layers learning later suffer.

Deep learning is a technique of machine learning that consists of many hierarchical layers to process the information. Deep learning offers human-like multi-layered processing in comparison with the shallow architecture. The basic idea of deep learning is to employ hierarchical processing using many layers of architecture. The layers of the architecture are arranged hierarchically. Each layer's input is provided to its adjacent layer after some pre-training. Most of the time, such pre-training of a selected layer is done in an unsupervised manner. Deep learning follows a distributed approach to manage big data. The approach assumes that the data are generated considering various factors, different times and various levels. Deep learning facilitates arrangement and processing of the data into different layers according to its time (occurrence), its level, or nature. Deep learning is often associated with artificial neural networks.

Nitish Srivastava et al. (2014) defines deep learning using a neural network in this way: 'deep neural networks contain multiple non-linear hidden layers and this makes them very expressive models that can learn very complicated relationships between their inputs and outputs. Many mundane applications such as natural speech, images, information retrieval and other human-like information processing applications may be benefited by employing deep leaning neural networks. For such applications, deep learning is suggested. Google is the pioneer of experimenting with deep learning, as initiated by the Stanford computer scientist Andrew Ng (now at Baidu as chief scientist). Experiment with Google's 'deep dream' images floating around when you are Googling!

There are three categories of deep learning architecture: (1) generative, (2) discriminative and (3) hybrid deep learning architectures. Architectures belonging to the generative category focus on pre-training of a layer in unsupervised manner. This approach eliminates the difficulty of training the lower level architectures, which rely on the previous layers. Each layer can be pre-trained and later included into the model for further general tuning and learning. Doing this resolves the problem of training neural network architecture with multiple layers and enables deep learning.

Neural network architecture may have discriminative processing ability by stacking the output of each layer with the original data or by various information combinations, thus forming deep learning architecture. According to Deng Li (2012), the discriminative model often considers neural network outputs as a conditional distribution over all possible label sequences for the given input sequence, which will be optimized further through an objective function. The hybrid architecture combines the properties of the generative and discriminative architecture.

Deep learning is helpful to effectively manage the great volume of complex data in a business. Many different models and algorithms have made it practically possible to implement deep learning in various domains. The possible applications of deep learning in the field of data science are given below.

- Natural language processing and natural query.
- Google's automatic statistician project.
- Intelligent web applications including searching and intelligent crawling.
- Image, speech and multi-media mining.
- Utilization of social network platforms for various activities.
- Development for resources and sustainable development, population information, governance (weather forecasting, infrastructural development, natural resources).
- Sensor web, agricultural information, decision support systems in domains such as forestry and fisheries.

5.4.5 Other Models of ANN

Besides the above-mentioned models of artificial neural network, there are other models available as follows.

5.4.5.1 Self-Organizing Map

A self-organizing map (SOM) is a kind of artificial neural network with the unsupervised learning paradigm. It is also known as a self-organizing feature map (SOFM). Historically, Finnish Professor Teuvo Kohonen (1982) has developed such self-organizing map; that is why it sometimes called a Kohonen map or network.

Like every type of artificial neural network, this SOM also consists of neurons. Along with every neuron, a weight factor and position information of the neuron on the map are associated. Generally, the neurons are arranged in a two-dimensional regular spacing in a grid format. Typically, SOM uses a neighbourhood function to preserve the topological properties of the input space.

The weights of the neurons are determined in a random manner as usual. This network also works in two phases: training and mapping. During training, example vectors are fed to the network and a competitive learning algorithm is applied. It is desirable to have example vectors similar to the expected actual mapping situation (as close as the actual expectations). When a sample training example is provided to the network, Euclidean distance to all weight vectors is computed. A neuron is identified whose weight factor matches with the input provided. This neuron is called the best matching unit (BMU). The weights of this best matching neuron and its neighbours are adjusted. The update formula for a neuron v with weight vector $Wv(s)$ is given as follows:

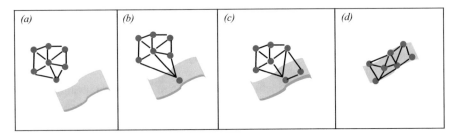

Fig. 5.9 Learning process in the Kohonen map (**a**) Network and training data (**b**) Initial training phase (**c**) Network converges towards training data (**d**) Network actually maps the training data

$$Wv\,(s+1) = Wv(s) + \Theta\,(u,\ v,\ s)\ \alpha(s)\,(D(t) - Wv(s)),$$

where s is the step index;
t an index into the training sample;
u is the index of the BMU for D(t);
$\alpha(s)$ is a monotonically decreasing learning coefficient;
$D(t)$ is the input vector;
$\Theta(u,\ v,\ s)$ is the neighbourhood function which gives the distance between the neuron u and the neuron v in step s. (Kohonen and Honkela 2007)

Eventually, the neighbourhood function shrinks with time in a stepwise manner. Earlier, when the neighbourhood is broad, the self-organizing takes place on the global scale; afterward it is converging to local estimates.

This process is repeated for each input vector for a number of cycles. Figure 5.9 shows the learning process in the Kohonen map.

SOMs are used for applications in domains such as meteorology and oceanography, deep learning and software project prioritization. Sometimes they are also used for optimization.

There are other models of artificial neural network such as simple recurrent network, fully recurrent network, Boltzmann machine and modular networks. A recurrent neural network has at least one feedback loop, which generally is the output of the network. It has been observed that all kind of biological neural networks are recurrent in nature. According to the structure and design of an artificial neural network, a learning algorithm is adopted. There exist so many learning algorithms; some use training examples and some learn even in the absence of training examples.

Such neural network algorithms are well defined and are being experimented within various domains. To implement the algorithms, there are tools and methodologies available if one is not interested in extending the effort to develop an application specific tool using the programming languages and packages.

5.4.6   Linear Regression and Neural Networks

The final layer has a particular role in a deep neural network of many layers. When dealing with labelled input, the output layer classifies each example, applying the most likely label. Each node on the output layer represents one label, and that node turns on or off according to the strength of the signal it receives from the previous layer's input and parameters.

Each output node gives two possible outcomes, the binary output values 0 or 1, since an input variable either deserves a label or does not. While neural networks working with labelled data give binary output, the input they receive is frequently continuous. To be precise, the signals that the network receives as input will span a range of values and include any number of metrics, depending on the problem it seeks to solve.

For example, a recommender engine has to make a binary decision about whether to serve an advertisement or not. But the input it bases its decision on could include how much a shopper has spent on 'Snapdeal' in the last 2 days, or how often that shopper visits the site.

So the output layer has to condense signals such as € 102 spent on footwear and eight visits to a website into a range between 0 and 1; i.e., a probability that a given input should be labelled or not.

The procedure used to convert continuous signals into binary output is called logistic regression. It computes the probability that a set of inputs match the label.

$$F(x) = \frac{1}{1 + e^{-x}}$$

For incessant inputs to be expressed as probabilities, they must output positive results, since there is no such thing as a negative probability. Therefore, the input is shown as the exponent of e in the denominator because exponents make results greater than zero. Further consider the relationship of e's exponent to the fraction $\frac{1}{1}$. As the input x that triggers a label grows, the expression e to the x shrinks toward zero, leaving us with the fraction $\frac{1}{1}$, or 100 %, which means we approach absolute certainty that the label applies. Input that correlates negatively with your output will have its value flipped by the negative sign on e's exponent, and as that negative signal grows, the quantity e to the x becomes larger, pushing the entire fraction ever closer to zero.

Now suppose that, instead of having x as the exponent, you have the sum of the products of all the weights and their corresponding inputs—the total signal passing through your network. That's what you're feeding into the logistic regression layer at the output layer of a neural network classifier. With this layer, we can set a decision threshold above which an example is labelled 1, and below which it is not.

Real World Case 3: Forecasting
In the travel industry, customer transactional data can be useful in the development of a forecasting model that accurately produces meaningful expectations. Regardless of whether a restaurant relies on moving average or time series forecasting algorithms, ANN can improve the statistical reliability of forecast modelling. Estimating in advance how much and when menu items will need to be prepared is critical to efficient food production management. ANN can provide a prediction of product usage by day given available sales data. In addition, knowing how much product was sold during any meal period can also be helpful in supporting an effective inventory refill system that minimizes the amount of capital tied up in stored products.

5.5   ANN Tools and Utilities

A tool is a process or device that is used to achieve a goal. Some tasks are not possible without using a proper tool such as a knife or saw (you cannot cut a wooden sheet using a hammer or a dynamite; you must have a saw!). Similarly, a very essential tool in implementing and automating the learning process supported by a neural network is computer and necessary software. In this list, programming languages such as java, C++, python and R come first. Besides the programming languages, there are various tools available, as discussed below.

- MatLab: The MatLab is a framework based on fourth generation programming language. MatLab is abbreviated from Matrix Laboratory and developed by MathWorks.[2] This framework is widely used in MATLAB, which is widely used by academicians, researchers, professional and learners. Depending on the application domain, the framework works in conjunction with various toolboxes such as robotics toolbox, impage processing toolbox, and machine learning toolbox.
- Java Neural Network Simulator (JavaNNS): JavaNNS[3] is a friendly interface developed on the programming language java to help in the design and implementation of neural network. It is developed at the Wilhelm-Schickard-Institute for Computer Science (WSI) in Tübingen, Germany.
- NeuroSolutions: NeuroSolutions[4] is a neural network software package on the Windows platform. It offers a graphical user interface using icons and menus to

---

[2]http://in.mathworks.com/

[3]http://www.ra.cs.uni-tuebingen.de/software/JavaNNS/

[4]http://neurosolutions.com/neurosolutions/

design artificial neural network with MS Office (Excel) interface. This tool is very popular in the group of the users who are comfortable with Microsoft Office pages.

- NeuroXL: This is another tool to define and experiment with artificial neural network using a Microsoft Excel interface.[5] It offers component such as NeuroXL Predictor, NeuroXL Clusterizer and OLSOFT Neural Network Library.
- FANNTool: This is a graphical user interface developed on the FANN library.[6] It is a user friendly and open source package to create artificial neural network.
- Weka: This is a compilation of techniques to design and implement data mining tasks. Besides neural network, it also supports implementation of various dada mining algorithms. As per the claim of the developer, the Weka 3.0 version is capable of supporting data pre-processing, classification, regression, clustering, association rules, and visualization.[7]
- Brain: This simulator is to design and implement artificial neural network. The Brain simulator is available in most of the platforms. It is user friendly, flexible and extensible. It is written in the Python programming language.
- Multiple Back-Propagation: Multiple Back-Propagation[8] is free software for implementing neural networks. As its name denotes, it supports multi-layer perceptron design with back propagation learning mechanism.
- NNDef Toolkit: This is also a free package for experimenting with artificial neural network.[9] It allows designing neural network using utilities namely NNDef Runtime Engine, NNDef Runtime Library, and NNDef Transformers and Sample Models.
- NuMap and NuClass: These products are developed by Neural Decision Lab.[10] This freeware supports neural network design, training, and validation.
- Joone Object Oriented Neural Engine: This is a framework written in Java to develop artificial neural network.[11] It consists of a core engine, a graphical user interface (GUI) editor and a distributed training environment. One can extend the framework by adding new modules to implement innovative algorithms.

5.6 Emotions Mining on Social Network Platform

In the modern era, the social network platform has become an essential mechanism of being connected with people. Due to unavailability of time and the high effort of reaching many people, it is nearly impossible to be in touch with friends,

---

[5]http://www.neuroxl.com/

[6]https://code.google.com/p/fanntool/

[7]http://www.cs.waikato.ac.nz/ml/weka/

[8]http://mbp.sourceforge.net/

[9]http://www.makhfi.com/nndef.htm

[10]http://www.neuraldl.com//Products.htm

[11]http://www.jooneworld.com/

relatives, colleagues and professionals. Using the information and communication technology advancements and availability of social network platforms such as Twitter, Facebook and LinkedIn; it is now possible to be social with many interesting people at the same time. It enables us to know about people, products and trends in our areas of interest.

Many data are available on such a platform. These data can be utilized in many useful applications. However, the problem is that these data are unstructured, voluminous and full of errors and emotions. Such really big data are difficult to acquire, clean and analysed, due to aforementioned reasons. This section introduces an experiment to effectively utilize these data for identifying emotions about a given item, and hence to promote the item.

5.6.1 Related Work on Emotions Mining

Social network platforms are emotionally rich environment with moods, ideas, interests, liking and disliking. To extract hidden knowledge about a given entity has been of prime interest of many researchers since initial phase of the twenty-first century, immediately after the popularization of social network platforms. Some early works were carried out by Peter Turney (2002); Bo Pang and Lillian Lee (2004) demonstrated applications of sentiment mining for products and films. Dmitri Roussinov and Leon Zhao (2003) developed an application to identify sense from a text message. Opinion mining was also experimented on, on the semantic web by Priti Srinivas Sajja and Rajendra Akerker (2012). Feature-based opinion mining has also been experimented on by Priti Srinivas Sajja (2011). Sentiment analysis is also useful for identifying abusive postings in Internet newsgroups (Spertus 1997). Emotions extraction is a field very near to sentiment analysis and opinion mining. Mining emotions from a test has also been experimented on by various authors (Dhawan et al. 2014; Neviarouskaya et al. 2010; Aman and Szpakowicz 2007 and Yang et al. 2009).

5.6.2 Broad Architecture

To mine the emotions from a given social network platform, artificial neural network is used. The broad architecture of the system for emotions mining is illustrated in Fig. 5.10.

The very first phase of the system is to acquire the text content from the selected social mining platform. For that, one should have proper access rights and enough social connections in order to retrieve text. As stated earlier, the text is full of errors and emotions, and lacks any specific structure. One may obtain help from the inbuilt

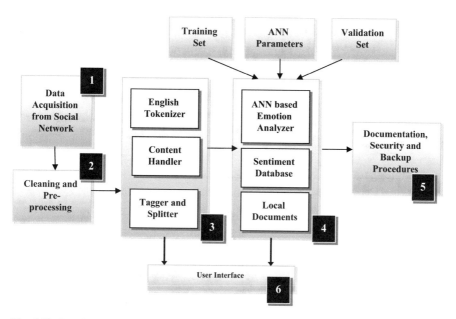

Fig. 5.10 Broad architecture of the emotions mining from social network platform

facility of the social network platform selected to collect all transactions between the connections. For example, Facebook allows you to backup a conversation using the gear icon available on the top right corner in one easy step.

Before we mine emotions from the material, the material must be cleaned. As we are considering only texts and icons to extract emotions, images and other multi-media content can be removed from the collected text. Other items that need to be removed from the collected content are keywords such as 'http' and 'www'. Also, select text components with high similarities of reflecting emotions such as 'I feel', 'I am', 'makes me', 'happy', 'sad' and 'like'. During this process, word stemming is necessary; that is, unhappy, sadness and sad are considered to be the same word. It is also necessary to remove all non-alpha numeric characters, and all text is converted into the lower case. All these processes are carried out in the second phase of the system.

Once data are cleaned and pre-processed, the icons and smiles used in the text are encoded. Furthermore, each word from the text is to be tokenized. This process is done in the third phase of the system. Besides the English language Tokenizer, there are two other functionalities that come into the picture in the third phase of the system. These functionalities are content handler and tagger and splitter. The main objective of the content handler is to simplify the text content as described above. Optionally, the content handler (as well as other components of the third phase) can be integrated with the second phase of the system, with cleaning and pre-processing.

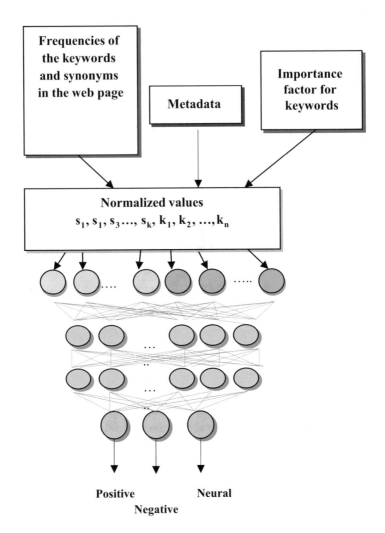

Fig. 5.11 Architecture of the system

5.6.3 Design of Neural Network

In the next phase, phase four, an artificial neural network is designed and trained with good quality samples. Figure 5.11 presents the architecture of the system, along with structure of the artificial neural network.

The tokenized keywords identified from the content with their associated frequencies are provided to the input layer of the neural network. Optionally, synonyms of the keywords and metadata can also be provided to the input layer of the neural network.

Table 5.3 Collection of keywords and synonyms with frequencies in a retrieved text

No.	Keywords	Frequency
Keyword 1	Happy	3
Keyword 2	Like	1
Keyword 3	Certainly	1
Keyword 4	Good	2
Keyword 5	Sad	2
....	......	....

Table 5.4 Normalized values to the input layer of the neural network

Keyword	K1	K2	K3	K4	K5
Frequency	3	1	1	2	2
Normalized value	0.75	0.65	0.65	0.50	0.50

The structure of the training vector encoded from the text retrieved encompasses extracted keywords along with their frequencies. Suppose the text contains phrases such as 'happy', 'good', or 'like' with high frequencies, then the text is considered to have positive emotions about an entity. If there are words showing negative emotions with high frequency, then the text is categorized as negative emotions. An example collection is given in Table 5.3.

From the aforementioned information a vector is formed, which is further multiplied with application of a specific weight factor supplied by the user and normalized before sending to the neural network. An example of such vector is given in Table 5.4.

Training requires such multiple vectors. Furthermore, output of the neural network is again a vector tagging the text into 'Positive', 'Neutral' or 'Negative' categories. The user interface asks for the user's choice before presenting the content. It is to be noted that the quality of the decision of the neural network is based on the training data provided.

5.7 Applications and Challenges

ANN can be useful to automatically learn from a large amount of data. It is good to have lot of data to make a neural network learn. This ability of an ANN to learn, to make tunings to its structure over time, is what makes it valuable in the field of data science. Pattern recognition, signal processing, control, time series prediction, and anomaly detection are selected usages of ANN in software today.

Furthermore, ANN can be applied to almost every problem where you have historical data and a need to build a model for that data. For example:

- Portfolio Management: Assign the assets in a portfolio in a way that maximizes return and minimizes risk.
- Credit Rating: Automatically assigning an enterprise's or individual's credit rating based on their financial situation.

- Medical Diagnosis: Supporting physicians with their diagnosis by analysing the reported symptoms and/or image data such as MRIs or X-rays.
- Process Modeling and Control: Creating an artificial neural network model for a physical plant then using that model to determine the best control settings for the plant.
- Machine Diagnostics: Detect when a machine has failed so that the system can automatically shut down the machine when this occurs.
- Intelligent Searching: An Internet search engine that provides the most relevant content and banner advertisements based on the user's previous behaviour.
- Fraud Detection: Detect unauthorized account activity by a person for which the account was not intended.
- Target Recognition: Military application that uses video and infrared image data to determine if an enemy target is present.
- Targeted Marketing: Finding the set of demographics which have the highest response rate for a particular marketing campaign.
- Financial Forecasting: Using the historical data of a security to predict the future movement of that security.

Here are some applications of artificial neural network in the field of data science.

- Convergence of high dimensional data to low dimensional data such as geo-physical data (Hinton and Salakhutdinov 2006; Rezaie et al. 2012) can be done with this technique. Further, the technique can be used to predict earthquakes (Reyes et al. 2013) and wind data forecasting too (Babu and Reddy 2012).
- Classifications of multi-media data, such as imagenet content classification (Krizhevsky et al. 2012; Smirnov et al. 2014).
- Speech recognition (Sainath et al. 2013) and face detection (Zhang and Zhang 2014).
- Financial data analysis such as stock market analysis (Hsieh et al. 2011) and sales forecasting (Choi et al. 2011) can also be candidate applications of the technology.

In recent days, deep learning models are able to make use of enormous amounts of data to mine and extract meaningful representations for classification and regression. However, deep learning poses some specific challenges in big data, including processing massive amounts of training data, learning from incremental streaming data, the scalability of deep model, learning speed.

Usually, learning from a large number of training samples provided by big data could obtain complex data representations at high levels of abstraction, which can be used to improve the accuracy of classification of the deep model. An evident challenge of deep learning is the numerous formats of training sample, including high dimensionality, massive unsupervised or unlabelled data, noisy and poor quality data, highly distributed input data and imbalanced input data. The current deep learning algorithms cannot adjust to train on such kinds of training samples, while how to deal with data diversity is a truly huge challenge to deep learning models.

Streaming data is one of the key characteristics of big data, which is large, fast moving, dispersed, unstructured and unmanageable. This type of data extensively exists in many areas of society, including web sites, blogs, news, videos, telephone records, data transfer, and fraud detection. One of the big challenges of learning meaningful information with deep learning models from streaming data is how to adapt deep learning methods to handle such incremental unmanageable streaming data.

The challenge in training speed of deep learning basically comes from two features: the large scale of the deep network and the massive amount of training sample provided by big data. It has been turned out that focus on models with a very large number of model parameters, which are able to extract more complicated features and improve the testing accuracy greatly, can become excessively computationally expensive and time consuming. Furthermore, training a deep model on huge number of training data is also time consuming and requires a large amount of compute cycles. Consequently, how to accelerate the training speed of a large model in big data with a powerful computing device, as well as distributed and parallel computing is a challenge.

To train on large-scale deep learning models faster, a key approach is to speed up the training process with distributed and parallel computing such as clusters and GPUs. Current approaches of parallel training include data parallel, model parallel and data-model parallel. But when training large-scale deep models, each of these will be low in efficiency for the sake of parameter synchronization, which needs frequent communications between different computing nodes. Moreover, the memory limitation of modern GPUs can also lead to scalability of deep networks. Here the challenge is how to optimize and balance workload computation and communication in large-scale deep learning networks.

Real World Case 4: Customer Relationship Management
Companies in the competitive market rely primarily on persistent profits that come from existing loyal customers. Therefore, customer relationship management (CRM) always concentrates on loyal customers who are the most fertile and reliable source of data for managerial decision making. This data reflects customers' actual individual product- or service-consuming behaviour. This kind of behavioural data can be used to evaluate customers' potential life-time value, to assess the risk that they will stop paying their invoices or will stop using any products or services, and to anticipate their future needs. An effective Customer Relationship Management (CRM) program can be a direct outcome of ANN applications. In order to effectively control customer churn, it is important to build a more effective and accurate customer churn prediction model. Statistical and ANN techniques have been utilized to construct churn prediction models. The ANN techniques can be

(continued)

used to discover interesting patterns or relationships in the data, and predict or classify the behaviour by fitting a model based on available data. In other words, it is an interdisciplinary field with a general objective of predicting outcomes and employing cutting-edge data processing algorithms to discover mainly hidden patterns, associations, anomalies, and/or structure from extensive data stored in data warehouses or other information repositories. The ability to enhance CRM given rapid accessibility of more comprehensive management information should lead to satisfied customers and improved sales performance. The ability to anticipate and influence consumer behaviour can provide a company with a competitive advantage. Having a signature item, for example, can be found to be a driver of improved relations while providing a product that customers do not perceive as having an equivalent elsewhere.

5.8   Concerns

One of the major issues with neural networks is that the models are rather complicated. The linear model is not only intuitively better, but also performs quantitatively better as well. But this leads to a noteworthy point about training and evaluating learning models. By building a very complex model, it's easy to entirely fit our data set. But when we evaluate such a complex model on new data, it performs very poorly. In other words, the model does not *generalize* well. This is known as *overfitting*, and it is one of the prime challenges that a data scientist must combat. This turn out to be major issue in deep learning, where neural networks have large numbers of layers containing many neurons. The number of connections in these models is enormous. As a result, overfitting is commonplace. One method of tackling overfitting is called *regularization*. Regularization modifies the objective function, which we minimize by adding additional terms that penalize large weights. The most usual kind of regularization is *L2 regularization*. It can be implemented by augmenting the error function with the squared magnitude of all weights in the neural network. Another common type of regularization is *L1 regularization. L1* regularization has the intriguing property of leading the weight vectors to become sparse during optimization (i.e., very close to exactly zero). In other words, neurons with *L1* regularization end up using only a small subset of their most important inputs and become quite resistant to noise in the inputs. Moreover, weight vectors from *L2* regularization are mostly diffuse, small numbers. *L1* regularization is very useful when you want to understand exactly which features are contributing to a decision. If this level of feature analysis is not mandatory, we can use *L2* regularization because it empirically performs better.

Another method for preventing overfitting is called *Dropout*. While training, dropout is implemented by only keeping a neuron active with some probability p (a hyper-parameter), or setting otherwise it to zero. This drives the network to be accurate even in the dearth of precise information. It averts the network from becoming too dependent on any one of neurons. It averts overfitting by offering a means of almost exponentially combining many different neural network architectures efficiently.

Finally, it can be said that, any industry where the accuracy of its predictions can make a significant financial impact to its business could benefit most from deploying Neural Network tools and techniques. For a company like Netflix, increasing the accuracy of movie recommendations from what they were doing before by 10 % might not be a huge deal. But for a company involved in any kind of algorithmic trading, an extra 10 % increase in the quality of certain decisions in their pipeline can make a really big difference to their bottom line. Oil and gas exploration is another example. But these are the obvious ones—these kinds of techniques can also be applied to self-driving cars, housekeeping robots, car driving robots, algorithmically generated and tailored to a person's unique gamer tastes, and so on.

5.9 Exercises

1. Concisely describe the different development phases of artificial neural networks and give representative examples for each phase.
2. Would it be beneficial to insert one bias neuron in each layer of a layer-based network, such as a feed forward network? Examine this in relation to the representation and implementation of the network.
3. Why are learning algorithms important to artificial neural networks? Justify.
4. Give possible applications of artificial neural networks for the following:

   • Decision support
   • Character recognition
   • Transaction processing

5. Describe the principle of neural computing and discuss its suitability to data science.
6. *(Project)* Deep learning has shown great performance on visual classification tasks and on object localization. Yet, applying deep neural networks to general visual data analytics and interpretation is still not fully achieved. Explore recent advances in state-of-the-art computer vision and machine learning theories, and study deep learning architectures to create an innovation in the field of visual data processing and analysis, including deep learning in the context of video understanding, particularly action recognition, activity analysis and pose estimation.

7. *(Project)* Implement an ANN model of your choice. Test the model on at least two different data sets and for a reasonable spread of the learning algorithm parameters. Analyse and discuss your results.

References

Ackley, D., Hinton, G., & Sejnowski, T. (1985). A learning algorithm for Boltzmann machines. *Cognitive Science, 9*(1), 147–169.

Aman, S., & Szpakowicz, S. (2007). Identifying expressions of emotion in text. In *Proceedings of 10th international conference on text, speech and dialogue* (pp. 196–205). Plzen: Springer.

Babu, C., & Reddy, B. (2012). Predictive data mining on average global temperature using variants of ARIMA models. In *International conference on advances in engineering, science and management* (pp. 256–260). Nagapattinam, India.

Carpenter, G., & Grossberg, S. (1988). The ART of adaptive pattern recognition by a self-organizing neural network. *IEEE Computer, 21*(3), 77–88.

Choi, T. M., Yu, Y., & Au, K. F. (2011). A hybrid SARIMA wavelet transform method for sales forecasting. *Decision Support Systems, 51*(1), 130–140.

Deng, L. (2012). A tutorial survey of architectures, algorithms, and applications for deep learning. *APSITA Transactions on Signal and Information Processing*, 1–29.

Dhawan, S., Singh, K., & Khanchi, V. (2014). A framework for polarity classification and emotion mining from text. *International Journal of Engineering and Computer Science, 3*(8), 7431–7436.

Fukushima, K. (1988). A neural network for visual pattern recognition. *IEEE Computer, 21*(3), 65–75.

Hinton, G. E., & Salakhutdinov, R. R. (2006). Reducing the dimensionality of data with neural networks. *Science, 313*(5786), 504–507.

Hopfield, J. J. (1982). Neural networks and physical systems with emergent collective computational abilities. *Proceedings of National Academy of Sciences of the USA, 79*(8), 2554–2558.

Hsieh, T. J., Hsiao, H. F., & Yeh, W. C. (2011). Forecasting stock markets using wavelet transforms and recurrent neural networks: An integrated system based on artificial bee colony algorithm. *Applied Soft Computing, 11*(2), 2510–2525.

Kohonen, T. (1982). Self-organized formation of topologically correct feature maps. *Biological Cybernetics, 43*(1), 59–69.

Kohonen, T. (1988). *Self-organization and associative memory*. New York: Springer.

Kohonen, T., & Honkela, T. (2007). Kohonen network, *2*(1), 1568.

Krizhevsky, A., Sutskever, I., & Hinton, G. (2012). ImageNet classification with deep convolutional neural networks. *Advances in Neural Information Processing Systems, 25*, 1106–1114.

McCulloch, W., & Pitts, W. (1943). A logical calculus of the ideas imminent in nervous activity. *Bulletin of Mathematical Biophysics, 5*, 115–133.

Minsky, M., & Papert, S. (1969). *Perceptrons*. Cambridge: MIT Press.

Neviarouskaya, A., Prendinger, H., & Ishizuka, M. (2010). EmoHeart: Conveying emotions in second life based on affect sensing from text. *Advances in Human-Computer Interaction*.

Pang, B., & Lee, L. (2004). A sentimental education: Sentiment analysis using subjectivity summarization based on minimum cuts. In *Proceedings of the association for computational linguistics* (pp. 271–278). Barcelona.

Reyes, J., Marales-Esteban, A., & Martnez-Ivarez, F. (2013). Neural networks to predict earthquakes in Chile. *Applications of Soft Computing, 13*(2), 1314–1328.

Rezaie, J., Sotrom, J., & Smorgrav, E. (2012). Reducing the dimensionality of geophysical data in conjunction with seismic history matching. In *74th EAGE conference and exhibition incorporating EUROPEC 2012*. Copenhagen, Denmark.

Rosenblatt, F. (1957). *The perceptron: A perceiving and recognizing automaton.* Buffalo: Cornel Aeronautical Laboratory.

Roussinov, D., & Zhao, J. L. (2003). Message sense maker: Engineering a tool set for customer relationship management. In *Proceedings of 36th annual Hawaii International Conference on System Sciences (HICSS).* Island of Hawaii: IEEE.

Rumelhart, D., & Zipser, D. (1985). Feature discovery by competitive learning. *Cognitive Science, 9*(1), 75–112.

Sainath, T. N., Kingsbury, B., Mohamed, A. R., Dahl, G. E., Saon, G., Soltau, H., & Beran, T. (2013). Improvements to deep convolutional neural networks for LVCSR. In *IEEE workshop on automatic speech recognition and understanding* (pp. 315–320). Olomouc, Czech Republic.

Sajja, P. S. (2011). Feature based opinion mining. *International Journal of Data Mining and Emerging Technologies, 1*(1), 8–13.

Sajja, P. S., & Akerkar, R. (2012). Mining sentiment using conversation ontology. In H. O. Patricia Ordóñez de Pablos (Ed.), *Advancing information management through semantic web concepts and ontologies* (pp. 302–315). Hershey: IGI Global Book Publishing.

Smirnov, E. A., Timoshenko, D. M., & Andrianov, S. N. (2014). Comparison of regularization methods for ImageNet classification with deep conveolutional neural networks. *AASRI Procedia, 6,* 89–94.

Spertus, E. (1997). Smokey: Automatic recognition of hostile messages. In *Proceedings of conference on innovative applications of artificial intelligence* (pp. 1058–1065). Menlo Park: AAAI Press.

Srivastava, N., Hinton, G., Krizhevsky, A., Sutskever, I., & Salakhutdinov, R. (2014). Dropout: A simple way to prevent neural networks from overfitting. *Journal of Machine Learning Research, 15,* 1929–1958.

Turney, P. (2002). Thumbs up or thumbs down? Semantic orientation applied to unsupervised classification of reviews. In *Proceedings of 40th annual meeting of the association for computational linguistics* (pp. 417–424). Philadelphia.

Yang, S., Shuchen, L., Ling, Z., Xiaodong, R., & Xiaolong, C. (2009). Emotion mining research on micro-blog. In *IEEE symposium on web society* (pp. 71–75). Lanzhou: IEEE.

Zhang, C., & Zhang, Z. (2014). Improving multiview face detection with multi-task deep convolutional neural networks. In *IEEE winter conference on applications of computer vision* (pp. 1036–1041). Steamboat Springs.

Chapter 6
Genetic Algorithms and Evolutionary Computing

6.1 Introduction

Evolutionary algorithms are inspired from the Nature's ability to evolve. Evolutionary algorithms are a component of evolutionary computing in the field of Artificial Intelligence. They are inspired from the biological evolution of random population by employing various modifying operations on the basic pattern of the candidates of the population. In nature, evolution through such modification happens in such a way that the next population will consist of members that are comparatively more fit to survive in the given situation. In a case where the modification results in poorer candidates, they cannot survive, and hence they will be automatically deselected from the population. Only the fit candidates will survive in the population. That is why such an approach is called the survival of the fittest approach. A broad outline of the approach is given as follows.

- Generate initial population by selecting random individuals.
- Apply one or more evaluation functions for the candidates.
- Select fit (good quality) candidates and push them in the next generation directly, if they are up to the mark.
- Select some strong candidates and modify them to generate even stronger candidates and push them to the next generation.
- Repeat the procedure until the population evolves towards a solution.

Evolutionary Algorithms is an umbrella term used to describe a consortium of various computational models/techniques following the basic principle of evolution. All the computational techniques that fall under the area of evolutionary algorithms apply various modifications to the population and preserve better candidates, and hence evolve the population. Techniques such as genetic algorithms, genetic programming, evolutionary programming, gene expression programming, evolution

© Springer International Publishing Switzerland 2016
R. Akerkar, P.S. Sajja, *Intelligent Techniques for Data Science*,
DOI 10.1007/978-3-319-29206-9_6

Table 6.1 Evolutionary algorithm techniques

Technique	Description
Genetic algorithm	It is the most popular technique that falls under evolutionary algorithms.
	The candidates are represented by strings over a finite set of alphabets, numbers or symbols. Traditionally binary number system is used.
	It is often used to solve optimization problems.
Genetic programming	The candidates within the population are in form of computer code. The code that can solve a given computational problem is considered to be a fit candidate.
	Often the individuals are represented through tree encoding strategy.
Evolutionary strategy	The candidates in populations are represented as real valued vectors. The techniques generally use self-adaptive mutation rates.
Evolutionary programming	This technique is very much similar to genetic programming. The difference lies in the structure of the computer code used as individuals. Here, the structure of computer code is fixed; however, parameters are allowed to evolve.
Gene expression programming	This technique also uses various computer codes as individuals in the population. Here, computer codes of different sizes are encoded into linear chromosomes of pre-defined length using tree data structure.
	The individuals represented using tree structures evolve by changing their sizes, shapes, and composition.
Differential evolution	Differential evolutions use a similar evolutionary approach for real numbers with specific mutation and crossover operators.
	Here, the candidates are represented in the form of vectors and emphasis is given on vector differences. This technique is basically suited for numerical optimization.
Neuro-evolution	This is similar to genetic programming; however, the difference is in the structure of the candidate. Here, the individuals of a population are artificial neural networks.
Learning classifier system	Here, the individuals of a population are classifiers represented in the form of rules or conditions. The rules can be encoded in binary (direct approach) or using artificial neural network or symbolic expression (indirect approach).
Human-based genetic algorithm	The technique is similar to genetic algorithm or genetic programming. However, the human is allowed to direct the evolutionary process. Often the human serves as the objective function and evaluates potential solutions.

strategy, differential evolution, neuro-evolution and learning classifier systems fall under the umbrella of evolutionary algorithms. Table 6.1 describes the techniques in brief.

For any evolutionary system, the fundamental components are encoding strategy using the individuals that are represented, modifying operators and evaluation functions. Modification of operators generate new candidates to form a new generation from which better candidates are selected. It is the evaluation function that preserves the good quality solutions within the populations. However, sometimes the weak candidates may survive and become parents as some of their characteristics are really strong. This way, one may claim that evolutionary algorithms are stochastic. The common characteristics of evolutionary algorithms are listed below.

- Evolutionary algorithms preserve population, modify their candidates and evolve stronger candidates.
- Evolutionary algorithms modify candidates by combining adorable characteristics of individual candidates to generate a new candidate that is stronger than the original candidates.
- Evolutionary algorithms are stochastic in nature.

6.2 Genetic Algorithms

Being a candidate of the evolutionary algorithm consortium, a genetic algorithm (GA) follows the basic principle of evolution as stated earlier (refer to Section 1, broad outline of evolutionary algorithm). The genetic algorithm is generally used for searching and optimization problems in an adaptive manner. The process of evolution is very similar to biological evolution and works according to the survival of the fittest principle introduced by the famous naturalist and geologist Charles Darwin. It is also called the principle of natural selection. In nature, It is observed that species and individuals compete for resources such as food, water, shelter and mates. Those who can succeed in these exercises will survive and can produce many offspring. Losers are not able to survive, they cannot produce offspring, and hence the gene pattern of weak individuals cannot be transferred. Only the genes from adaptive and fit individuals will be passed to the next generations. The offspring in many cases have better combinations of genes from their parents and become much fitter than the parents. Generation by generation, the species will become stronger. See the popular story of giraffes in Fig. 6.1.

A typical Genetic algorithm is given at Fig. 6.2 and illustrated in Fig. 6.3.

The basic principles of genetic algorithms were first explored by John Holland (1975). Since then, many other scientists have been working on the technique to solve problems in various domains. Genetic algorithms are adaptive and robust in nature. They are useful while the processing voluminous data, which are complex and interrelated. Genetic algorithms are useful and efficient when:

- The search space is large, complex, or poorly understood.
- Domain knowledge is scarce or expert knowledge is difficult to encode to narrow the search space.
- No mathematical analysis is available.

Furthermore, genetic algorithms are random algorithms—the course taken by the algorithm is determined by random numbers. In particular, if you ask a random

It is believed that the early giraffes were not tall like today's giraffes. They were much shorter in comparison. The Charles Darwin survival of the fittest theory explains how all giraffes at present have long neck. He explained that, at the time of drought, during scarce situation, some of the giraffes with long necks, could managed to eat more leaves from the far branches could survive. Giraffes with short height (and neck) could not survive and died. Most of the survived giraffes had long neck and good height. The next generation produced by these tall giraffes was having many tall giraffes as the parents are tall. In such a way the whole generation went through the genetic modification, parents' genes of good height were transformed to the children. By accident, if there were any short giraffes, they could not survive; as they are not fit for the environment. By this way the species of giraffes has become tall.

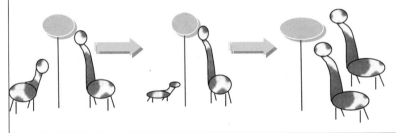

Tall giraffes live and breed & short giraffes give up … ends up with all tall giraffes.

In the case of giraffe, the characteristic of good height is very important to survive, and giraffes who could maintain good height could survive. This shows that the Nature will support the solutions with the desired characteristics. Solutions which are fit and strong can only survive.

Fig. 6.1 Tale of giraffes evolution

algorithm to optimize the same problem twice in precisely the same manner, you will get two different answers. Sometimes you'd like to get the same answer to the same question. If so, this is against random algorithms—though it would be possible to always use the same seed for a random number generator.

The common principles of the genetic algorithm are discussed in the next section.

Begin

- Generate initial population of individuals (known or random).

- Encode the individuals by taking suitable encoding stream. Also determine string length of encoded string for an individual.

- Repeat the steps given below till the desired fitness level of the individuals is reached or improvement is not further observed or up to predefined number of iterations (say maximum 100).

    - Apply genetic operators and generate new population.

    - Evaluate fitness of the individuals.

    - Select stronger individuals and remove poor ones from the new population.

End

Fig. 6.2 Typical genetic algorithm

6.3 Basic Principles of Genetic Algorithms

Before application of genetic algorithms to a problem, the individuals from the problem domain must be represented in the form of genes of predetermined length. As genetic algorithms follow the concept of evolutionary algorithms, this starts with a population encompassing randomly selected (if not known) and encoded individuals. These individuals are frequently modified and evaluated for their fitness, until a satisfactory solution is reached. For these, there needs to be an encoding strategy (for representation of characteristics of individuals), genetic operators such as mutation and crossover (to modify individuals), and fitness functions (to push good quality individuals into the next generation). This section describes these basic principles with necessary examples.

6.3.1 Encoding Individuals

Individuals selected from the domain of interest need to be encoded in the form of genes. Each gene or set of genes represents a characteristic. A set of genes is also commonly known as a gene string, phenotype or genotype. However, a genotype is the set of genes in the string which is responsible for a particular characteristic, whereas a phenotype is the physical expression, or characteristics. Each individual is encoded and represented in the form of a series of genes/genotype. That is why

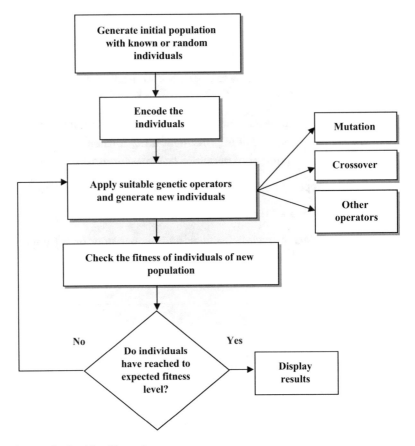

Fig. 6.3 Genetic algorithm illustration

Individual X	1	0	0	1	1	0	0	1
Individual Y	0	1	1	1	0	0	1	0

Fig. 6.4 Binary encoding

an individual is identified as a chromosome. John Holland (1975) has used binary values to encode individuals. In binary encoding, each individual is represented in the form of a binary number (a number system having only two symbols, 0 and 1, with base 2). Each binary digit (bit) represents a gene.

Figure 6.4 shows an example of individuals X and Y encoded in a binary number system with a string length of 8.

After encoding, initialization of the population is done. While initializing a population, random individuals from the search space are chosen or it can be filled with known values.

6.3.2   Mutation

Mutation is basically a bit flipping operation. It is a method of changing a gene with another valid value, with the help of a mutation probability, say $P_{Mutation}$. It is done as follows. A number between 0 and 1 is chosen at random. If the random number is smaller than $P_{Mutation}$, then the mutation is done at the current bit, otherwise the bit is not altered. For binary encoding, a mutation means to flip one or more randomly chosen bits from 1 to 0 or from 0 to 1. Figure 6.5 shows two original individuals X and Y, as shown in Fig. 6.4, and a mutation operation on them. On the individual represented by label X, a two-site mutation is done at position 3 and position 5. On the individual represented by Y, a two-site mutation is done at position 1 and position 7.

6.3.3   Crossover

Using mutation alone is not sufficient; rather, it is just like a random walk through the search space. Other genetic operators must be used along with mutation. Crossover is another such operator. Crossover selects substrings of genes of the same length from parent individuals (often called mates) from the same point, replaces them, and generates a new individual. The crossover point can be selected randomly. This reproduction operator proceeds as follows.

* The reproduction operator selects at random a pair of two individual strings.
* A cross-site is selected at random within the string length.
* The position values are swapped between two strings following the cross-site.

Individual X	1	0	0	1	1	0	0	1
New Individual X	1	0	1	1	0	0	0	1
Individual Y	0	1	1	1	0	0	1	0
New Individual Y	1	1	1	1	0	0	0	0

Fig. 6.5 Two-site mutation on individual X and individual Y

Individual X	1	0	0	1	1	0	0	1
Individual Y	0	1	1	1	0	0	1	0
Individual P	1	1	1	1	1	0	0	1
Individual Q	0	0	0	1	0	0	1	0

Fig. 6.6 Crossover on binary encoding with string length 3, position 2

Individual X	1	0	0	1	1	0	0	1
Individual Y	0	1	1	1	0	0	1	0
Individual P	1	1	1	1	1	0	1	0
Individual Q	0	0	0	1	0	0	0	1

Fig. 6.7 Crossover on binary encoding with string length 3, position 2 and string length 2, position 7

Such operation results in two new individuals. Crossover selects some desirable characteristics from one parent and other characteristics from another parent and creates a new individual by taking the best of both (just like a father's height and a mother's skin in a child!). Figure 6.6 illustrates the process of crossover.

Simultaneous crossover can also be done at multiple sites, as shown in Fig. 6.7.

6.3.4 Fitness Function

Fitness function plays an important role in determining the quality of individuals and hence the quality of the population providing the end solutions. It is the control that is set on the individuals for allowing them into the next generation: individuals that are fit as per the fitness function that only would be pushed further. It is observed that tighter and better fitness function results in high quality individuals in a generation. A fitness function must be well defined in order to allow individuals with similar fitness to be close and together. The fitness function must lead the evolution towards valid and good individuals. The fitness function also works as a penalty function for poor individuals. Some penalty functions also provide completion cost. The completion cost is factor showing expected cost of converting an invalid individual into a valid one. In some cases, instead of exact fitness or penalty function, approximate functions are also used.

6.3.5 Selection

A selection operator chooses good individuals for further operations. It generally copies good strings, but does not create new ones. Individual solutions are selected through a well-defined fitness function. Many definitions of selection operators have been in application. John Holland's (1975) *fitness-proportionate selection* is a pioneer. In the method defined by the Holland, individuals are selected with a probability proportional to their relative fitness. Such fitness-proportionate selection is also known as Roulette-wheel selection. If f_i is the fitness of individual i in the population, its probability of being selected is $P_i = f_i / \sum f_i$, where i takes values from 1 to N, where N is the number of individuals in the population. The name 'Roulette-wheel solution' comes from the analogy of a typical Roulette wheel in casino games. Each individual represents a pocket on the wheel; the size of the pockets is proportionate to the probability of the individual being selected. Instead of probability, percentages can also be considered. Selecting N individuals from the population is equivalent to playing N games on the Roulette wheel, as each candidate is drawn independently. That is, selection is done by rotating the wheel a number of times equal to the population size. Figure 6.8 illustrates the typical Roulette wheel selection.

Because of the probability function utilized here, in the Roulette wheel selection method, there is a chance that stronger individuals with a higher fitness value may be deselected. With this approach, another chance is that weaker individuals may be selected. It is beneficial sometimes; as such weaker solutions may have a few genes that are really strong. A strong part of the weaker individual may then be further considered for crossover function.

Another method that eliminates a fixed percentage of the weakest candidate is the tournament selection method. Tournament selection works as follows.

Fig. 6.8 Roulette wheel selection

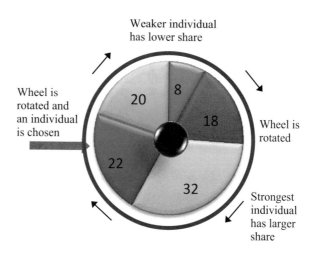

Table 6.2 Tournament selection

Begin
Determine size of tournament, say K
Create mating pool M
Selected individual set A
Determine probability P
Select K individuals from the population at random
Repeat for K individuals
{Evaluate fitness and store in A}
Select best from A and insert it to the mating pool M with probability P
Select second best from A and insert it to the mating pool M with probability $P (1 - P)$
Select third best from A and insert it to the mating pool M with probability $P (1 - P) (1 - P)$
End

- A pair of individuals are selected randomly and evaluated for their fitness.
- The stronger individual is inserted into a mating pool.
- This process is repeated until the mating pool is filled completely.
- Modifying operators and probability based fitness functions are applied to these individuals.

Here, we are selecting an individual from a pair, hence it is called binary tournament. At a time many individuals (say n) can also be selected for comparison, this is called larger tournament with size n. If the tournament size is larger, strong individuals have better chances to be selected. To make better individuals win, one can consider a probability factor along with comparison through fitness function, where an individual will win the tournament with probability p (usually above average, i.e., > 0.5). Table 6.2 displays a broad outline of the tournament selection.

While selecting individuals, some researchers adopt a methodology for selection such that no individual of the current population will be duplicated. Some select n new individuals and delete n individuals and keep the population in fixed size and steady state status. Another possibility is to deselect all and add any number of new individuals.

Using selection alone will tend to fill the population with copies of the best individual from the population. However, the selection cannot introduce new individuals into the population. For inclusion of new individuals, one has to have the help of operators such as mutation and crossover.

6.3.6 Other Encoding Strategies

So far we have seen binary encoding strategy for individuals with fixed length bit strings, which is a popular encoding strategy. One can also use symbols, alphabets, and tree structures for encoding. Figure 6.9 presents mutation and crossover operations on trees X and Y, respectively.

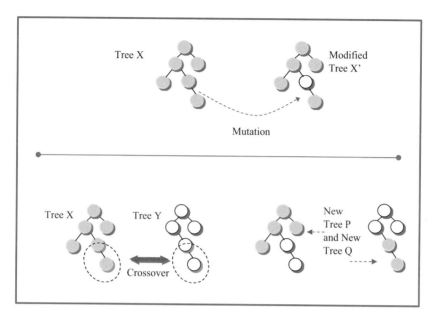

Fig. 6.9 Tree encoding and operations

Generally, tree encoding would be used in evolving programs in genetic programming as well as gene expression programming. The syntax of the instructions can be represented in the form of trees on which operations like selection, mutation, and crossover are carried out. The resulting offspring are evaluated against a fitness function.

Real World Case 1: Airplane Design Optimization
A primary airplane design can be realized by means of genetic algorithms (GA). The airplane key parameters are mapped into a chromosome-like string. These include the wing, tail and fuselage geometry, thrust requirements and operating parameters. GA operators are performed on a population of such strings and natural selection is expected to occur. The design performance is obtained by using the airplane range as the fitness function. GA cannot only solve design issues, but also project them forward to analyze limitations and possible point failures in the future so these can be avoided.

6.4   Example of Function Optimization using Genetic Algorithm

This example demonstrates use of the genetic algorithm for function optimization. Consider a function $f(x, y) = x * y$ for maximization within the interval of [0, 30]. A genetic algorithm finds values of x and y (an order pair (x,y)) that produce maximum values of the function $f(x, y)$ from the given interval [0, 30].

Encoding
Individuals are encoded using binary digits (bits) with length 5. The possible values of x and y are between 0 and 30. These numbers can be represented into five bits maximum. Taking more binary bits results in larger numbers than 30 and fall out of the specified interval for values x and y. Hence, such numbers are invalid.

Fitness Function
The fitness function definition is already provided in the problem definition. The function definition $(x*y)$ will be the fitness function. The individuals x and y are added to check the fitness.

The initial population is selected randomly as denoted in Table 6.3.

On the initial population, fitness of individuals are calculated, according to which a Roulette wheel is designed, which is shown in Fig. 6.10.

Selection, mutation and crossover operations are carried out on the basis of the Roulette wheel count shown in Fig. 6.10. The resulting individuals are denoted in Table 6.4.

Table 6.3 Initial population

Sr. no. of individual	Individuals (First 5 bits for x and last 5 bits for y)	Decimal values of x and y	Fitness $f(x, y) = x*y$ (Value in Decimal)	Roulette wheel selection count
1	0101101010	10,11	110	1
2	1100001100	24,12	288	2
3	0100100101	09,05	45	0
4	0101100111	11,07	77	1

Fig. 6.10 Roulette wheel selection for the example

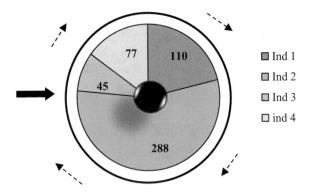

- Ind 1
- Ind 2
- Ind 3
- ind 4

Table 6.4 Operations on the population

Sr. no. of selected individual	Individuals (First 5 bits for x and last 5 bits for y)	New individual after mutation or crossover	Operation site (and string length)	Fitness $f(x, y) = x*y$ (value in decimal)
1	01**0**11 **0**1010	01**1**11 **1**1010	Mutation at position 3 and 6	390
2	11000 **0**1100	11000 **1**1100	Mutation at position 6	672
3	110**00** 01100	110**11** 01100	Crossover with mate 4, site 4, length 2	324
4	010**11** 00111	010**00** 00111	Crossover with mate 3, site 4, length 2	56

Table 6.5 Second generation

Sr. no. of individual (new)	Individuals (first 5 bits for x and last 5 bits for y)	Decimal value of x and y	Fitness $f(x, y) = x*y$ (value in decimal)	Roulette wheel selection count
1	01111 11010	15, 26	390	1
2	11000 11100	24, 28	672	2
3	11011 01100	27, 12	324	0
4	01000 00111	8, 7	56	1

One can see in Table 6.4 that individual 3 is removed from the population and individual 2 is duplicated. This decision is taken from the experiments of the Roulette wheel. The count of the roulette wheel is denoted in Table 6.3. It is obvious that individual 2 is the strongest (fittest) in the first generation, as shown in Table 6.3. The poorest individual is the individual 3; hence, it is removed from the next generation. After selection of stronger individuals in the next generation, new individuals are created by applying the mutation and crossover function. Table 6.5 shows the second generation.

One can observe an increase in fitness values for majority of the individuals in Table 6.5, illustrating the second generation of the population. The Roulette wheel for the second generation individuals is shown in Fig. 6.11.

Continuing the evolution, one can reach to the optimum solution. Here, we are have the advantage of knowing the solution (through application of the traditional method) and we may smartly select operations (mutation and crossover) in such a way that we can reach the solution in a few generations. In the case of real problems, where traditional solutions are not available, one can set the maximum number of iterations of evolution or the minimum improvement factor in the solution. If there is no significant improvement after evolution, and stagnation is achieved, the process may be stopped. Ideally, the genetic algorithm should evolve in such a way that all its individuals are successively showing better fitness.

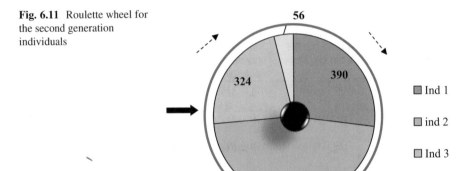

Fig. 6.11 Roulette wheel for the second generation individuals

6.5   Schemata and Schema Theorem

Schema is a template that identifies a subset of strings with similarities at certain string positions. The notion of schema was introduced by John Holland (1975). A schema is a template defined over the alphabet $\{0, 1, *\}$, which describes a pattern of bit strings in the search space $\{0, 1\}^L$ (the set of bits strings of length L). For each of the L bit positions, the template either specifies the value at that position (1 or 0), or indicates by the symbol * (referred to as 'don't care' or wildcard symbol) that either value is allowed.

The schema **1 0 * * 1 0 *** is a template for the following strings:

$$1\ 0\ 11\ 1\ 0\ 1$$
$$1\ 0\ 00\ 1\ 0\ 0$$
$$1\ 0\ 01\ 1\ 0\ 1$$
$$1\ 0\ 01\ 1\ 0\ 0$$
$$1\ 0\ 10\ 1\ 0\ 1$$
$$1\ 0\ 10\ 1\ 0\ 0$$

6.5.1   Instance, Defined Bits, and Order of Schema

A bit string x that matches a schema's **S** pattern is said to be an instance of **S**. In a schema, 1s (ones) and 0s (zeroes) are referred to as defined bits. The total number of defined bits in the schema is considered as an order of a schema. The defining length of a schema is the distance between the leftmost and rightmost defined bit in the string, as shown in Table 6.6.

Table 6.6 Order, length and instances of schema

Schema	Order	Length	Instances
1 1 * * 1 1 0 *	5	6	1 1 1 1 1 1 0 1
			1 1 0 1 1 1 0 0
* * * * 1 1 0 *	3	2	11 10 1 1 0 0
			1011 1 1 0 1
**0*0*1	3	4	1101011
			1001001
1**0*1	3	5	100011
			111011

6.5.2   Importance of Schema

Schema in one way provides an explanation of how a genetic algorithm works. The basic idea is to evaluate fitness of a schema (that is, group of individuals represented by the schema), instead of applying the fitness function to randomly generated individuals from the whole population. Each schema represents a set of individuals following the common pattern. Evaluating the pattern reduces the burden of testing each individual following the schema. It can be said that the fitness of the schema is the fitness of the individuals following the pattern specified by the schema. This is an important and efficient way to manage a big search space and impart generalization in the evolution process. This way, schemas will be helpful in searching good regions of the search space. A schema that provides above-average results has greater chances of providing better elements in terms of fitness. As per the Building Block Hypothesis (Hayes, 2007), the genetic algorithm initially detects biases towards higher fitness in some low-order schemas (with a small number of defined bits) and over time, detects the same in high-order schemas. John Holland (1975) has also suggested that the observed best schemas will, on average, be allocated an exponentially increasing number of samples in the next generation. Holland also suggested that an individual may satisfy more than one schema, some of which are stronger and some of which are weaker. Indirectly, the manipulation is done on these schema in such a way that stronger schema will survive. This is called an implicit parallelism.

6.6   Application Specific Genetic Operators

Typical genetic operators such as crossover and mutation may not be suitable for many applications. Consider the travelling sales person problem. The traveling salesperson problem is about finding an optimum path of a tour from a given set of cities so that each city is visited only once and the total distance travelled is minimized. To encode the possible routes as individuals, either city initials (alphabets) or symbols are used. One can also use numbers. If there are five different cities to be travelled once with the minimum effort, the possible paths using numbers as encoding strategy can be given as follows:

Route 1 : (1 2 3 4 5)
Route 2 : (2 3 4 5 1)

The standard mutation and crossover operators here create illegal/invalid solutions/plans. See the following example, in which Route 1 undergoes a mutation operation at location 1. City 1 can be replaced with one of the remaining cities (2, 3, 4, or 5). This results in an illegal plan.

Original Route 1 : (1 2 3 4 5)
Mutated Route 1 : (2 2 3 4 5) (invalid)

Similarly, a crossover (at location 2 with string length 3) also results in illegal solutions, as follows:

Route 1 : (1 2 3 4 5)
Route 2 : (2 3 4 5 1)

New Offspring Route 3 : (1 3 4 5 5) (invalid)
New Offspring Route 4 : (2 2 3 4 1) (invalid)

To avoid such illegal solutions, one can do following.

• Try a different representation (encoding scheme).
• Design a special crossover operator that is application specific and generates valid offspring.
• Design a penalty function that removes illegal individuals by giving high penalty (say, negative or low fitness).

In the case of the travelling salesperson problem, fitness function is very straightforward and rigid. It is not advisable to change the fitness function. Encoding strategy is also very limited in this case. Here, one can consider the new design of genetic operators that are suitable to this application and generate valid individuals.

With the help of the edge recombination technique, this problem can be solved. This technique works as follows.

• Create a population with finite number of legal tours as the initial generation.
• Create an adjacency table for each city in all routes. The adjacency table contains all cities along with all their possible neighbours in all the routes from the population.
• Generate new individuals by recombining the genes used in the parent individuals as follows:

   – Select the parent at random and assign its first gene as the first element in the new child.
   – Select the second element for the child as follows: If there is an adjacency common to both parents, choose that element to be the next one in the child's permutation; if there is an unused adjacency available from a parent, choose it. If these two options fail, make a random selection.
   – Select the remaining elements in order by repeating step

• Push the new individuals in the new population in order to evaluate its fitness.

6.6.1   Application of the Recombination Operator: Example

Consider five different cities labelled 1, 2, 3, 4, and 5 for a typical traveling salesperson problem (TSP). The two randomly generated individuals are as follows:

Route 1 : (1 2 3 4 5)

Route 2 : (2 3 4 5 1)

The new adjacency list can be given as follows:

Key	Adjacency list
1	2,5
2	1,3,3
3	2,4,2,4
4	3,5,3,5
5	4,4,1

With the previous algorithm, a new offspring can be generated, with City 3 as random starting point. Let us append the starting city in the new child route, called Route 3.

Route 3 : 3,

From city 3, next popular destination is either city 2 or 4 as per the adjacency table. Let us select city 2. The city 2 is appended at the route 3.

Route 3 : 3, 2,

From city 2, popular destinations are city 3 and 1. Since city 3 is already traversed, city 1 is selected and appended in to the Route 3.

Route 3 : 3, 2, 1,

From city 1, popular destinations are city 2 and 5. Since city 2 is already traversed, city 5 is selected and appended in to the Route 3.

Route 3 : 3, 2, 1, 5,

It is obvious that city 4 is not yet visited; hence we may append the city 4 directly. Otherwise, city 5's neighbours are city 4 and city 1. From the adjacency table, one can see that city 4 is comparatively more popular than city 1. City 1 is already visited, hence we select city 4. After appending city 4 in the new route, the new route will be as follows.

Route 3 : 3, 2, 1, 5, 4

Route 3 is a valid route as it lists each city once only. If the distances between cities are provided, one can also calculate the cost of the route. In our case, it is the fitness function.

Real World Case 2: Shipment Routing
Recent applications of a GA, such as the *Traveling Salesperson Problem*, can be used to plan the most efficient routes and scheduling for travel planners, traffic routers and even logistics companies. The main objectives are to find the shortest routes for traveling, the timing to avoid traffic delays, and to include pickup loads and deliveries along the way. See section 6.6 for further details.

6.7 Evolutionary Programming

Evolutionary Programming is inspired by the theory of evolution by means of natural selection. Precisely, the technique is inspired by macro-level or the species-level process of evolution (phenotype, hereditary, variation), and is not concerned with the genetic mechanisms of evolution (genome, chromosomes, genes, alleles). Evolutionary programming (EP) was originally developed by L. J. Fogel et al. (1966) for the evolution of finite state machines using a limited symbolic alphabet encoding. Fogel focused on the use of an evolutionary process for the development of control systems using finite state machine (FSM) representations. Fogel's early work elaborated the approach, focusing on the evolution of state machines for the prediction of symbols in time series data.

Evolutionary programming usually uses the mutation operator to create new candidate solutions from existing candidate solutions. The crossover operator that is used in some other evolutionary algorithms is not employed in evolutionary programming. Evolutionary programming is concerned with the linkage between parent and child candidate solutions and is not concerned with surrogates for genetic mechanisms.

Fogel broadened EP to encode real numbers, thus offering a tool for variable optimization. Individuals in the EP comprise a string of real numbers, as in evolution strategies (ESs). EP differs from GAs and ESs in that there is no recombination operator. Evolution is exclusively dependent on the mutation operator, which uses a Gaussian probability distribution to perturb each variable. The standard deviations correspond to the square root of a linear transform of the parents' fitness score (the user is required to parametrize this transform).

An EP outline can be given as:

1. A current population of μ individuals is randomly initialized.
2. Fitness scores are assigned to each of the μ individuals.
3. The mutation operator is applied to each of the μ individuals in the current population to produce μ offspring.
4. Fitness scores are assigned to the μ offspring.
5. A new population of size μ is created from the μ parents and the μ offspring using tournament selection.
6. If the termination conditions are satisfied, exit; otherwise, go to step 3.

6.8 Applications of GA in Healthcare

Genetic algorithms are also considered as search and scheduling heuristics. As stated earlier, in this chapter, when a problem domain is too large and traditional solutions are too difficult to implement, genetic algorithms are most suitable. Genetic algorithms can generate a good number of possible solutions and use them to find the best solutions with the specific fitness function. Data science activities include data identification and acquisition, data curation, data analytics and data visualization. These activities deal with large amounts of data that do not follow uniform structure. Here, genetic algorithms help in identifying cluster of data, managing the data, scheduling and processing data, and helping other activities of data science in many different ways. Here are some reasons why genetic algorithms are useful for managing data science-related activities.

- Genetic algorithms are useful to manage non-linearity in the search space.
- Genetic algorithms are good in searching from large and complex search spaces using heuristics; furthermore, genetic algorithms can perform global searches efficiently on such search spaces.
- Genetic algorithms are robust and adaptive in nature.
- Genetic algorithms can be integrated with other technologies such as artificial neural network and fuzzy logic; i.e., they are extensible in nature and easy to hybridize.
- Genetic algorithms are remarkably noise tolerant and evolving in nature.

6.8.1 Case of Healthcare

Healthcare is defined as caring about one's heath through proper and timely diagnosis, treatment and prevention of a disease or any uncomfortable health related situation in human beings. To practice healthcare at different levels, many organized systems work together. Being a multi-disciplinary field affecting a mass of people, healthcare practice generates a large amount of data, which must be handled with care in order to improve efficiency, reduce cost and increase availability. The area is simultaneously most important and complex. It is important, because it deals with the basic health of mankind, and complex because of its volume, its unstructured nature and variety of content. This is an area that involves more complicated data in large volume than many other industries. Furthermore, data that are the base of problem solving and decision making also have relative importance and are dynamic in nature. For example, in the health care domain, some viruses become resistant to an antibiotic after a certain period. This leads to the requirement of designing new antibiotic drugs. For this, a long history of many patients treated with similar antibiotics must be studied in order to determine the pattern of virus behaviour. Knowing such patterns would be helpful in designing new, powerful and effective antibiotic drugs. However, a new antibiotic drug will soon become ineffective as viruses become resistant to this new drug also. Further analysis and pattern-finding exercises need to be done, obviously more complex than the previous exercise. Day by day, the procedure becomes more complex and requires dedicated approaches such as genetic algorithms. This shows that with a large pool of complex data, the time has come to utilize data science techniques for healthcare-related data. Healthcare practitioners, insurers, researchers, and healthcare facilitators can be benefited by the use of advanced and latest trend applications in the field of data science.

As stated, the large pool of data generated through healthcare practices can be handled with data science. There are so many activities where healthcare data needs to be managed by handling its sheer volume, complexity and diversity. Data science techniques are helpful in facilitating cost effective data handling for decision making, diagnosing, resource management and care delivery. Data science activities such as data acquisition, data pre-processing and curation, data analytics and data visualization can be helpful in not only managing the data, but also in discovery of knowledge and improving the quality of services provided to patients, who are the ultimate end user of the facility.

Typically, healthcare information includes information related to the patients, doctors and other experts, resources, drugs and insurance claims. Information about these components is not available in a straightforward distinct manner, but in the form of complex mixture. Furthermore, the information is multi-media. It can be partially digitalized and stored in the form of multi-media and partially manual documents. Health informatics can be further subdivided into various domains, such as image informatics, neuro-informatics, clinical informatics, public health informatics, and bioinformatics.

Evolutionary algorithms or genetic algorithms can be useful for managing patient databases for various applications. To demonstrate use of genetic algorithms, a case of automatic scheduling of patients is discussed here. Automatic scheduling of patients is possible for a limited domain; however, in domains with wide scope, it is a very complex problem to automate. Here, one has to take care of patient's emergency, doctor's availability, infrastructure and resources availability as well as insurance-related issues. Hard computing methods are very rigid in nature and hence may not be suitable for applications. Genetic algorithms' capability of handling large search space can be utilized here. This approach may not offer any perfect or best solution; however, it offers good and acceptable solutions, which can be practically possible.

6.8.2   Patients Scheduling System Using Genetic Algorithm

Genetic algorithm, as discussed earlier in this chapter, starts with initial population, with randomly selected individual for the domains. Later on, the initial population undergoes fitness testing. A few elements may be removed, which are poor in terms of fitness, and some good candidates are preserved. Later, to generate more candidate solutions, operators such as mutation and crossover can be applied. This procedure is repeated until the satisfactorily strong candidate is evolved.

Let us consider information related to the task. As patients are at the centre of all the activities, the very first thing to come to mind is information related to patients. Information such as a patient's identification number, name, contact information, category of healthcare service required, and other personal information such as blood group and age, is basic information associated with a set of patients. Similarly, information about infrastructure such as branch or specialty (section) of the hospital, operation theatre, room allocated, ICU facility availed, nurses allocated, and medicines used are also required. Information about doctors, outside experts, supporting staff and medical insurance also plays an important role. The scheduling of surgery may depend on a patient's present conditions, availability of doctors and anaesthetists, and even on medical policy verification. Figure 6.12 illustrates the complexity of the scenario.

To authenticate the schedule once it is proposed, some common validations that can be used in these situations are as follows:

- No patient can undergo many activities at a time. Rather, typical sets of activities are well defined.
- Every resource can be used once at a time. That is, a unit of an operation theatre must be utilized at a time for a single surgery. Similarly, the disposable resources can be used only once at any time.
- Each activity can be performed once.
- Each activity can be performed in a specific order.

The candidate solutions (individuals) must be checked for the above constraints. However, satisfying many such validations, a candidate solution does not become

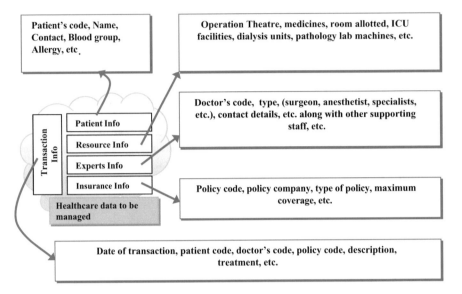

Fig. 6.12 Possible data in the application

the optimum solution. It has to survive the fitness function designed. A better idea is to apply the constraints at the time of encoding a possible solution. Various constraints, as mentioned above, are collected in order to design a well-defined constraints set, which can be used to validate the solutions suggested by the algorithm.

The overall design of the solution is illustrated as shown in Fig. 6.13.

The following section describes the activities shown in Fig. 6.13 with necessary details.

6.8.3 Encoding of Candidates

The encoding of an individual is done in such a way that it must represent necessary and important characteristics of an individual and simultaneously offer the ability to carry out genetic operations such as mutation and crossover. After applying crossover and mutation, again the derived solution must be checked against the constraint set to the validity of the solution. The derived solution must be feasible. Here, the individual solution is represented in multiple arrays. Each array represents the time characteristic, resource characteristic, and expert characteristic. Instead of taking multiple arrays, a single multi-dimensional array will also serve the purpose. Work done by Vili Podgorelec and Peter Kokol (1997) uses such a multi-dimensional array. An example of an encoding scheme is given at Fig. 6.14.

It is very clear from Fig. 6.14 that in a given time slot, a resource and expert can be used once only; hence leaving no chance of conflicts. Here in the T_1 time slot,

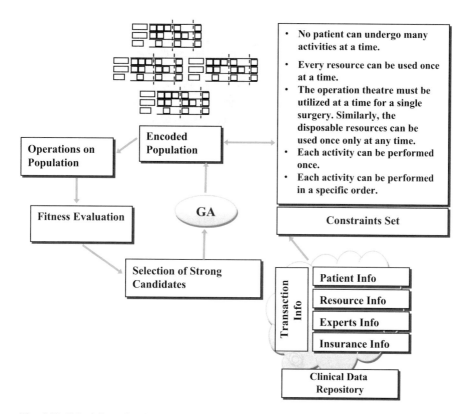

Fig. 6.13 Scheduling of patients' activities using genetic algorithm

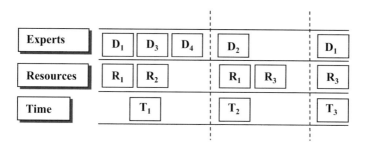

Fig. 6.14 Example encoding of individuals

resources R_1 and R_2 are used by expert doctors D_1, D_3 and D_4. It is not possible to intersect the resources R_1 and R_2 or experts in a given time period T_1. This encoding can be further improved by adding one more layer at an appropriate place, which manages added attributes in the encoding of individuals. For example, medical insurances status or any special requirement can be added, along with resources and experts.

The initial population contains many such randomly selected individuals. Prior to addition of an individual in the initial population, it must be checked for its validity. For that, the constraint set is applied. Any invalid candidates from the solution are removed from the population. Once valid candidates of the initial population are finalized, selection, mutation and crossover operations are performed to generate more candidates. Applying genetic operators such as mutation and crossover may result in an increased (or decreased) number of candidates in a population. The modified population is now called the new generation.

6.8.4   Operations on Population

This section describes the possible operations, such as selection, mutation and crossover, on the encoded individuals.

6.8.4.1   Selection

Selection can be done by ranking the individual schedules. To select an appropriate schedule, some of the following criteria can be considered.

- Time taken to complete overall activities.
- Idle time of one or more resources.
- Waiting time for patients.
- Consecutive schedules for a given expert.

For each candidate of a population, fitness score is evaluated using a function of the above parameters. A schedule must have as much as possible low waiting time; i.e., fewer chances of making a resource or expert idle while simultaneously leaving enough space for them to act comfortable. Using Roulette wheel or tournament selection, the candidates are removed or preserved in the population.

6.8.4.2   Mutation

Mutation requires only one candidate at a time. One candidate from the population is considered and some resources and/or experts are changed. This generates a new individual representing a novel schedule. This must be tested against the constraint set for its validity. Some resources and experts are restricted for mutation operations. Figure 6.15 illustrates the mutation operation on schedules.

In the example mentioned in Fig. 6.15, the mutation operation is done in such a way that resource 2 (labelled as R3) will not be in use consecutively.

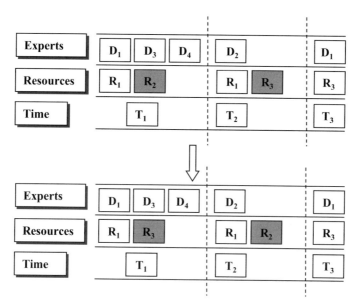

Fig. 6.15 Mutation operation on encoded individuals

6.8.4.3   Crossover

In crossover of resources, experts can also be possible in order to generate new individuals. The crossover function takes two individuals and interchanges the same number of resources and experts to generate resources. It is advisable to check the validity of the resulting individuals after applying the crossover function.

After a sufficient number of crossover and mutation operations, the population might have reached its saturation level. Any further operation might result in poorer candidates or show no improvement. Here, one can stop the process of evolution.

Real World Case 3: Environmental Monitoring
The objective is to perform measurements for pollution in city environments. In particular, outdoor environmental measurements for temperature, humidity and air pollution are included. Furthermore, some sensors are used to obtain such information. Mobile sensors on vehicles will be also utilized. A genetic algorithm can be adapted to the actual conditions of environmental problems, and then it can be used in environmental monitoring and environmental quality assessment.

6.8.5   Other Applications

Besides the healthcare domain, there are other domains where the genetic algorithms are used effectively. Some of the application areas are listed below:

- On the platform of Internet, there are lots of data available; much of these data are not directly useful. Consider the email correspondence between senior employees of an organization. Such a conversation contains high quality knowledge about the business. However, it may have some irrelevant information about sentiments and current trends. Further, the emails all differ in structure as far as content is considered. Other fields such as sender, receiver and subject of the mail are structured, but the mail content is in free form. Besides email, there are social network platforms, blogs and other websites also available on the same platform. Emails may refer to one of these in its content. Genetic algorithms can be helpful in managing these data by designing appropriate clustering algorithms. Collections of such data sets are available at Stanford University (Stanford Large Network Dataset Collection)[1] and at CMU,[2] which is an Enron Email Dataset prepared by the CALO (A Cognitive Assistant that Learns and Organizes) Project.

- Similar (to the above-mentioned one) applications of a genetic algorithm can be citation network analysis, collaborative platform network (such as Wikipedia) analysis, the Amazon networks, and online communities. Clustering, scheduling, analysis and visualization can be done using genetic algorithms on these domains.

- Just as clustering of knowledge or clustering of users as mentioned in previous example can occur, clustering can also be done on the Internet platform for various applications such as twitter, mail, chat, blog and other domains, and e-commerce users can also be clustered into various groups. E-commerce shopping agents/guide programmes may use such knowledge in proactively suggesting new products and assign in overall shopping. Furthermore, product feedback, promotion, product price optimization, and suggestions can also be manipulated. Even if their data are not on the Web, large and data-centric companies may also utilize genetic algorithms for analytics.

- Data mining and web mining are generic applications where the genetic algorithms can be utilized to mine complex hidden patterns on the data warehouses and the Web. For example, from large warehouses (which may be distributed in nature) or from voluminous repository of data, genetic algorithms can mine useful patterns or classification rules. Similarly, genetic algorithms can also be useful in learning concepts for large repositories. Such work is done by Jing Gao, et al. (2008).

- Finance and investment-related data are also complicated, as many parameters affect them simultaneously and continuously. Genetic algorithms may help in forecasting via the learning of complex patterns within the domain.

---

[1] http://snap.stanford.edu/data/

[2] http://www.cs.cmu.edu/~enron/

- In the area of automotive and engineering design, genetic algorithms can be used to find the optimum combination of best materials and best engineering design to provide faster, fuel efficient, safe and lightweight vehicles. Similar evolving hardware design concepts can also be used for the engineering data domain.
- Telecommunication data are also complex enough to handle. Genetic algorithms can be used to search for the best path and efficient routing for telephone calls. Genetic algorithms can also be utilized to optimize placement and routing of cell towers for best coverage and ease of switching. Similar applications can be designed in the fields of tourism and traffic manipulation.
- For computer gaming, even if categorized as formal tasks category because of well-defined rules (e.g., chess) future actions of the game can be suggested by the application of genetic algorithms.

Besides the above-mentioned applications in various domains, genetic algorithms may be used in scheduling, clustering, prediction, forecasting, and searching in general, by applying a human-like heuristic approach to solve challenging problems in the most efficient way.

Real World Case 4: Gene Expression Profiling
The development of microarray technology for taking snapshots of the genes being expressed in a cell or group of cells has been an advantage to medical research. These profiles can, for instance, distinguish between cells that are actively dividing, or show how the cells react to a particular treatment. Generic algorithms are being developed to make analysis of gene expression profiles much faster and simpler. This helps to classify what genes play a part in many diseases, and can assist in identifying genetic causes for the development of diseases. This is a key step in personalized medicine.

6.9 Exercises

1. Give a brief overview of a general evolutionary algorithm. Also, what are the similarities and differences between genetic algorithms and evolution strategies?
2. Explain the role of parameters of genetic algorithm that control mutation and crossover.
3. Given the following parents, P_1 and P_2, and the template T

P_1	A	B	C	D	E	F	G	H	I	J
P_2	E	F	J	H	B	C	I	A	D	G
T	1	0	1	1	0	0	0	1	0	1

Show how the following crossover operators work:

- uniform crossover
- order-based crossover

with regards to genetic algorithms.

Please note that uniform crossover uses a fixed mixing ratio between two parents to enable the parent chromosomes to contribute at the gene level rather than the segment level. For example, the mixing ratio is fixed to 0.5; then the offspring will have exactly half of the genes from one parent and the remaining half from the second parent. One may choose the crossover point randomly.

4. State two aspects of a genetic algorithm's design that would cause a population to fast converge. Refer to the two fundamental stages of genetic algorithms (selection and reproduction) in your answer.

5. Explain how you would solve a travelling salesperson problem using evolutionary algorithms. Illustrate your answer with suitable examples and diagrams.

6. Describe your own idea about linking an evolutionary algorithm to any machine learning method.

7. (*Project*) When we are trying to create an efficient portfolio of stocks, we must consider some important factors. The problem is that the evaluation contains several qualitative factors, which causes most approximations to go off track. Explore a genetic algorithm approach to portfolio evaluation. By using a set of fitness heuristics over a population of stock portfolios, the objective is to find a portfolio that has a high expected return over investment.

8. (*Project*) Investigate different ways to evolve neural networks. Start with comparing techniques of creating neural networks using genetic programming (GP). For instance, the function set in the GP can be utilized to create a directed graph containing the inputs, the output and any number of intermediate nodes.

9. (*Project*) Online social networks have generated great expectations in the context of their business value. Promotional campaigns implemented in web-based social networks are growing in popularity due to a rising number of users in online communities. In order to optimize a marketing campaign, explore hybrid predictive models based on clustering algorithm, genetic algorithms and decision trees.

References

Fogel, L. J., Owens, A. J., & Walsh, M. J. (1966). *Artificial intelligence through simulated evolution*. New York: Wiley.

Gao, J., Ding, B., Fan, W., Han, J., & Yu, P. (2008). Classifying data streams with skewed class distributions and concept drifts. *IEEE Internet Computing, Special Issue on Data Stream Management*, pp. 37–49.

Hayes, G. (2007, October 9). *Genetic algorithm and genetic programming*. Retrieved October 27, 2015, from http://www.inf.ed.ac.uk/teaching/courses/gagp/slides07/gagplect6.pdf

Holland, J. H. (1975). *Adaptation in natural and artificial systems*. Cambridge: The MIT Press.

Podgorelec, V., & Kokol, P. (1997). Genetic algorithm based system for patient scheduling in highly constrained situations. *Journal of Medical Systems, 21*, 417–447.

Chapter 7
Other Metaheuristics and Classification Approaches

7.1 Introduction

This chapter considers some of the effective metaheuristics and classification techniques that have been applicable in intelligent data analytics. Firstly, meta-heuristics approaches such as adaptive memory procedures and swarm intelligence are discussed, and then classification approaches such as case-based reasoning and rough sets are presented.

Metaheuristic algorithms have been applied to data mining tasks. Competitive metaheuristic methods are able to handle rule, tree and prototype induction, neural networks synthesis, fuzzy logic learning and so on. Also, the inherent parallel kind of metaheuristics makes them ideal for attempting large-scale data science problems. However, to accomplish a pertinent role in data science, metaheuristics need to have the capacity of processing big data in a realistic time frame, to efficiently use the exceptional computer power available at the present time. Fundamentally, all *heuristics* induce a pattern whose present state depends on the sequence of past states, and thus include an implicit form of memory. The explicit use of memory structures constitutes the core of a large number of intelligent solution methods. These methods focus on exploiting a set of strategic memory designs. The importance of high performance algorithms for attempting challenging optimization problems cannot be understated, and in several cases, the proper methods are metaheuristics. When designing a metaheuristic, it is desirable to have it be conceptually simple and effective.

Classification systems play an important role in business decision-making tasks, by classifying the available information based on some criteria. A variety of statistical methods and heuristics from artificial intelligence literature have been used in the classification tasks. Many of these methods have also been applied to different decision-making scenarios such as business failure prediction, portfolio management, and debt risk assessment. In case-based reasoning (CBR), the solution methodology is a process of comparison and evaluation of current needs with

© Springer International Publishing Switzerland 2016
R. Akerkar, P.S. Sajja, *Intelligent Techniques for Data Science*,
DOI 10.1007/978-3-319-29206-9_7

existing situations. However, the 'rules' approach is firmly grounded in cause and effect derivation of the reasons for doing specific tasks given a certain situation. Rough set theory is an approach concerned with the analysis and modelling of classification and decision problems involving vague, imprecise, uncertain, or incomplete information.

The methods discussed in this chapter have been successfully applied in many real life problems, including medicine, pharmacology, engineering, banking, finances, market analysis, environment management and others.

7.2 Adaptive Memory Procedure

Metaheuristics contains various algorithmic approaches, such as genetic algorithms, ant systems, adaptive memory procedures and scatter search. These are iterative techniques that use a central memory where information is collected during the search process.

The popular hybrid evolutionary heuristic is probably the genetic local search algorithm that combines a standard local search with standard genetic algorithm. A more recent hybrid metaheuristic is the adaptive memory procedure. An important principle behind adaptive memory procedure (AMP) is that good solutions may be constructed by combining different components of other good solutions. A memory-containing component of visited solutions is kept. Periodically, a new solution is constructed using the data in the memory and improved by a local search procedure. The improved solution is then used to update the memory.

(Taillard et al. 2001) characterized adaptive memory procedures as those that exploit a memory structure to obtain a solution. Specifically, they identified the following characteristics in these methods:

- A set of solutions or a special data structure that aggregates the particularities of the solutions produced by the search is memorized.
- A provisory solution is constructed using the data in memory.
- The provisory solution is improved using a greedy algorithm or a more sophisticated heuristic.
- The new solution is added to memory or is used to update the data structure, which memorizes the search history.

Over time, various uses of memory strategies have also become incorporated into a variety of other metaheuristics.

7.2.1 Tabu Search

Tabu search (TS) is a metaheuristic that guides a local heuristic search procedure to explore the solution space beyond local optimality (Gendreau 2002). One of the main components of TS is that it uses adaptive memory, which creates more flexible search behaviour. TS is based on the assumption that problem solving must incorporate adaptive memory and responsive exploration (Glover and Laguna 1997; Hertz and de Werra 1991). The TS technique is recently becoming the preference for designing solution procedures for hard combinatorial optimization problems. TS has also been used to create hybrid procedures with other heuristic and algorithmic methods, to provide improved solutions to problems in scheduling, sequencing, resource allocation, investment planning, telecommunications and many other areas.

In order to understand TS, first we must look at the structure of a local search (LS) algorithm.

The LS procedure begins with an *initial solution*, which is iteratively improved using *neighbourhood search* and *selection*. In each iteration, a set of *candidate solutions* is generated by making small modifications to the existing solution. The best candidate is then chosen as the new solution. The use of several candidates leads the search towards the highest improvement in the optimization function value. The search is iterated a fixed number of iterations or until a stopping criterion is met.

In an LS procedure, the following design aspects must be considered:

- Depiction of a solution: determines the data structures that are to be modified
- Neighbourhood function: defines the way the new solutions are generated
- Search strategy: establishes the way the next solution is chosen among the candidates.

Hill-climbing is a special case of LS where the new solutions are generated so that they are always better than or equal to the previous solution. Hill-climbing always finds the nearest local maximum (or minimum) and cannot improve anymore after that. The algorithm makes only *uphill* moves (i.e., application of neighbourhood function). Problem-specific knowledge is applied to design the neighbourhood function. It is also possible that there is no neighbourhood, but the move is a deterministic modification of the current solution that is known to improve the solution.

Tabu search (*TS*) is a variant of the traditional local search, using suboptimal moves but in a deterministic manner. It uses a *tabu list* of previous solutions (or moves), and in this way, prevents the search from returning to solutions that have been recently visited. This forces the search into new directions.

Real World Case 1: Forecasting Patient Demand for Efficient Hospital Recruitment

Nowadays, by incorporating all relevant and available data and by applying proven optimization techniques, super-speciality hospitals can both preserve their operating margins and enhance patient care. These hospitals often use data analytics solutions to precisely predict the inflow of patients and to determine whether or not the wait times have met the targets set by the

(continued)

hospital. If the wait times exceed the target, they run the simulation again until the target is met. In doing so, the tool is able to calculate the optimum number of nurses for each block (say, 5 hours per block). Then the tool feeds that number, as well as work patterns, approved holidays, preferences, workload per week, and other constraints, into the optimization models, placing nurses into shifts. Here the model can be developed, using a combination of tabu search and method(s) for approximating the global optimum, to find the best fit, minimizing the clashes that appear out of the many constraints that occur in a usual timetable.

The basic principle of TS is to pursue LS whenever it encounters a local optimum by allowing non-improving moves; returning back to previously visited solutions is prevented by the use of *memories*, called *tabu lists*, that record the recent history of the search, a key idea that can be linked to artificial intelligence concepts.

Recency-based memory is a memory structure used in TS implementations. This memory structure keeps track of solutions attributes that have changed during the recent past. To exploit this memory, selected attributes that occur in solutions recently visited are labelled tabu-active, and solutions that contain tabu-active elements, or particular combinations of these attributes, are those that become tabu. The use of recency and frequency memory in TS fulfils the function of preventing searching processes from loop.

One important trend in TS is hybridization, i.e., using TS in conjunction with other solution approaches such as genetic algorithms.

7.2.2 Scatter Search

Scatter search (SS) can be considered as an evolutionary or population-based algorithm that constructs solutions by combining others. The purpose of SS methodology is to enable the implementation of solution procedures that can derive new solutions from combined elements (Laguna and Mart'I 2003).

SS methodology is very flexible, since each of its elements can be implemented in a variety of ways and degrees of sophistication. The scatter search template serves as the main reference for most of the SS implementations to date.

In this subsection, we present a basic design to implement SS based on the 'five-method template'. The enhanced features of SS are related to the way these five methods are implemented. Figure 7.1 illustrates interaction among the five methods and highlights the central role of the reference set. The sophistication comes from the implementation of the SS methods instead of the decision to include or exclude certain elements (as in the case of TS or other metaheuristics).

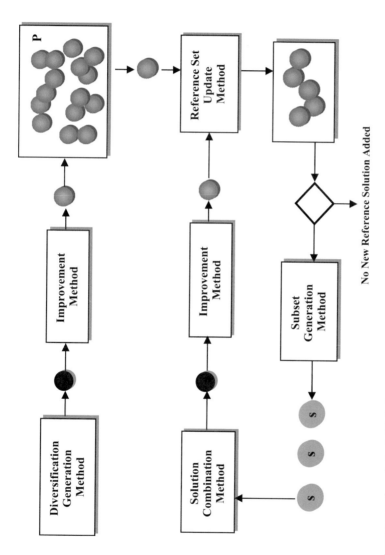

Fig. 7.1 Schematic representation of a basic SS design

The 'five-method template' for implementing SS is as follows:

1. A diversification generation approach to generate a collection of diverse trial solutions, using an arbitrary trial solution as an input.
2. An improvement method to convert a trial solution into one or more enhanced trial solutions. Neither the input nor the output solutions are required to be feasible. If no improvement of the input trial solution results, the enhanced solution is considered to be the same as the input solution.
3. A reference set update method to create and preserve a reference set comprising the best solutions found, organized to provide efficient accessing by other parts of the method. Solutions gain membership to the reference set according to their quality or their diversity.
4. A subset generation method to operate on the reference set, to generate several subsets of its solutions as a basis for creating combined solutions.
5. A solution combination method to convert a given subset of solutions produced by the Subset Generation Method into one or more combined solution vectors.

Unlike a *population* in genetic algorithms, the reference set of solutions in scatter search tends to be small. We have seen in Chap. 6 that in genetic algorithm, two solutions are randomly chosen from the population, and a 'crossover' or combination mechanism is applied to generate one or more offspring. A typical population size in a genetic algorithm consists of 100 elements, which are randomly sampled to create combinations. In contrast, a scatter search chooses two or more elements of the reference set in a systematic way with the purpose of creating new solutions. Since the combination process considers at least all pairs of solutions in the reference set, there is a practical need for keeping the cardinality of the set small. Typically, the reference set in a scatter search has 20 solutions or less. In general, if the reference set consists of b solutions, the procedure examines approximately $\frac{(3b-7)b}{2}$ combinations of four different types. The basic type consists of combining two solutions; the next type combines three solutions, and so on and so forth. Limiting the scope of the search to a selective group of combination types can be used as a mechanism for controlling the number of possible combinations in a given reference set.

In a nutshell, genetic algorithm approaches are predicated on the idea of choosing parents randomly to produce offspring, and further on introducing randomization to determine which components of the parents should be combined. By contrast, the SS approach does not emphasize randomization, particularly in the sense of being indifferent to choices among alternatives. Instead, the approach is designed to incorporate strategic responses, both deterministic and probabilistic, that take account of evaluations and history. SS focuses on generating relevant outcomes without losing the ability to produce diverse solutions, due to the way the generation process is implemented.

7.2.3 Path Relinking

Path relinking is a search technique that aims to explore the search space or path between a given set (usually two) of good solutions. The objective is to generate a set of new solutions in between the good solutions from the set.

Features that have been added to Scatter Search are captured in the Path Relinking framework. From a spatial orientation, the process of generating linear combinations of a set of reference solutions may be characterized as generating paths between and beyond these solutions, where solutions on such paths also serve as sources for generating additional paths. This leads to a bigger notion of the meaning of creating combinations of solutions. To generate the desired paths, it is mandatory to select moves that perform: Upon starting from an initiating solution, the moves must progressively introduce attributes contributed by a guiding solution or reduce the distance between attributes of the initiating and guiding solutions.

The roles of the initiating and guiding solutions are interchangeable; each solution can also be induced to move simultaneously toward the other as a way of generating combinations. The incorporation of attributes from elite parents in partially or fully constructed solutions was foreshadowed by another aspect of scatter search, embodied in an accompanying proposal to assign preferred values to subsets of consistent and strongly determined variables. The theme is to isolate assignments that frequently or influentially occur in high quality solutions, and then to introduce compatible subsets of these assignments into other solutions that are generated or amended by heuristic procedures.

Multi-parent path generation possibilities emerge in Path Relinking by considering the combined attributes provided by a set of guiding solutions, where these attributes are weighted to determine which moves are given higher priority. The generation of such paths in neighbourhood space characteristically 'relinks' previous points in ways not achieved in the previous search history, hence giving the approach its name.

Example: Let us consider two permutations of five tasks: $\eta_1 = \{a, b, e, c, d\}$, $\eta_2 = \{b, c, d, a, e\}$. We have the set of interchange movements to transform η_1 into η_2 like (a,1,4), which means that the task a placed on position 1 of η_1 will be placed on position 4 of η_1, which is the position of the task a in η_2.

At the same time, task c in η_1 (which is in position 4 of η_1) will be placed in position 1, i.e., interchanged with task 1 in η_1. The set of movements to transform η_1 into η_2 are showed in Table 7.1.

Table 7.1 Interchange movements of the path relinking to transform η_1 into η_2

Movements	Permutation
	(a,b,e,c,d) $= \eta_1$
(a,1,4)	(c,b,e,a,d)
(b,2,1)	(b,c,e,a,d)
(e,3,5)	(b,c,d,a,e) $= \eta_2$

With respect to the selection of the two individuals to carry out the path relinking, we can select them from the current population or we can select two individuals from a pool of elite solutions.

7.3 Swarm Intelligence

Swarm intelligence (SI) metaheuristics is the collective behaviour of (natural or artificial) decentralized, self-organized systems. SI systems are typically made up of a population of simple agents interacting locally with one another and with their environment (Kennedy and Eberhart 2001). The agents follow very simple rules, and although there is no centralized control structure dictating how individual agents should behave, local, and to a certain degree random, interactions between such agents lead to the emergence of 'intelligent' global behaviour, unknown to the individual agents. Natural examples of SI include ant colonies, bird flocking, animal herding, bacterial growth, and fish schooling.

In swarm intelligence, useful information can be obtained from the competition and cooperation of individuals. Generally, there are two kinds of approaches that apply swarm intelligence as data mining techniques. The first category consists of techniques where individuals of a swarm move through a solution space and search for solution(s) for the data mining task. This is a search approach; the swarm intelligence is applied to optimize the data mining technique, e.g., the parameter tuning. In the second category, swarms move data instances that are placed on a low-dimensional feature space in order to come to a suitable clustering or low-dimensional mapping solution of the data. This is a data organizing approach; the swarm intelligence is directly applied to the data samples, e.g., dimensionality reduction of the data.

In general, swarm intelligence is employed in data mining to solve single objective and multi-objective problems. Based on the two characters of particle swarm, the self-cognitive and social learning, the particle swarm has been utilized in data clustering techniques, document clustering, variable weighting in clustering high-dimensional data, semi-supervised learning based text categorization, and Web data mining. In a swarm intelligence algorithm, several solutions exist at the same time.

Swarm intelligence has been widely applied to solve stationary and dynamical optimization problems in the presence of a wide range of uncertainties. Generally, uncertainties in optimized problems can be divided into the following categories.

1. The fitness function or the processed data is noisy.
2. The design variables and the environmental parameters may change after optimization, and the quality of the obtained optimal solution should be robust against environmental changes or deviations from the optimal point.
3. The fitness function is approximated, such as surrogate-based fitness evaluations, which means that the fitness function suffers from approximation errors.

4. The optimum in the problem space may change over time. The algorithm should be able to track the optimum continuously.
5. The target of optimization may change over time. The demand of optimization may adjust to the dynamic environment.

In all these cases, additional measures must be taken so that swarm intelligence algorithms are still able to satisfactorily solve dynamic problems.

7.3.1 Ant Colony Optimization

The Ant Colony Optimization (ACO) metaheuristic is used for finding solutions and near solutions to intractable discrete optimization problems (Teodorovic´ and Dell'Orco 2005). The problem to be solved is modelled as a search for a minimum cost path in a graph. Artificial ants walk the graph, with each path corresponding to a potential solution to the problem. The behaviour of the ants is inspired by that of real ants: they deposit pheromone (a chemical substance) on the path in a quantity proportional to the quality of the solution represented by that path; they resolve choices between competing destinations probabilistically, where the probabilities are proportional to the pheromone gathered on previous iterations. This indirect form of communication, known as stigmergy (a mechanism of indirect coordination between agents or actions), intensifies the search around the most promising parts of the search space. On the other hand, there is also a degree of pheromone evaporation, which allows some past history to be forgotten, to diversify the search to new areas of the search space. The trade-off between intensification and diversification is influenced by modifying the values of parameters.

The features that ants in an ACO share with their real counterparts are:

- The ability to react to their local environment, to *smell* which of several trails is most attractive.
- A limited amount of vision, to *see* which of the trails in their immediate neighbourhood is shortest.
- The ability to *deposit* pheromone.

This means that any one ant does not have an overview of the entire network. It is restricted to sensing its immediate environment. This makes the ability to solve global optimization problems more exciting.

7.3.2 Artificial Bee Colony Algorithm

The artificial bee colony (ABC) algorithm is a swarm intelligence procedure based on the natural food searching behaviour of bees. Tereshko developed a model of foraging behaviour of a honeybee colony based on reaction–diffusion equations (Tereshko and Loengarov 2005). The main constituents of Tereshko's model are below:

1. Food Sources: In order to select a food source, a forager bee evaluates several properties related to the food source, such as its closeness to the hive, its richness of energy, the taste of its nectar, and the ease or difficulty of extracting this energy.
2. Employed foragers: An employed forager is active at a specific food source which she is currently exploiting. She carries information about this specific source and shares it with other bees waiting in the hive. The information includes the distance, the direction and the profitability of the food source.
3. Unemployed foragers: A forager bee that looks for a food source to exploit is called unemployed. It can be either a scout who searches the environment randomly or an onlooker who tries to find a food source by means of the information given by the employed bee.

In an ABC algorithm, the position of a food source represents a possible solution to the optimization problem and the nectar amount of a food source corresponds to the quality (fitness) of the associated solution.

In a D-dimensional search space, each solution (S_{xy}) is represented as

$$S_{xy} = \{S_{x1}, S_{x2}, \ldots, S_{xD}\}$$

Here, $x = 1, \ldots, SP$ is the index for solutions of a population and $y = 1, .., D$ is the optimization parameters index.

The probability value, which is based on the individuals' fitness value to summation of fitness values of all food sources and decides whether a particular food source has potential to obtain the status of a new food source, is determined as;

$$P_g = \frac{f_g}{\sum f_g}$$

where f_g and P_g are the fitness and probability of the food source g, respectively.

Once sharing the information between the existing onlookers and employed bees, in the case of a higher fitness than that of the previous one, the position of the new food source is calculated as follows:

$$V_{xy}(n + 1) = S_{xy}(n) + \left[\varphi_n \times \left(S_{xy}(n) - S_{zy}(n)\right)\right]$$

where $z = 1, 2, .., SP$ is a randomly selected index and has to be different from x. $S_{xy}(n)$ is the food source position at nth iteration, whereas $V_{xy}(n + 1)$ is its modified position in $(n + 1)^{th}$ iteration. φ_n is a random number in the range of $[-1, 1]$. The parameter S_{xy} is set to meet the acceptable value and is modified as

$$S_{xy} = S_{min}^y + ran(0, 1)\left(S_{max}^y - S_{min}^y\right)$$

In this equation, S_{max}^y and S_{min}^y are the maximum and minimum yth parameter values.

Although the employed and scout bees effectively exploit and explore the solution space, the original design of the onlooker bee's movement only considers the relation between the employed bee food source, which is decided by Roulette wheel selection, and a food source having been selected randomly. This consideration reduces the exploration capacity and thus induces premature convergence. In addition, the position-updating factor utilizes a random number generator which shows a tendency to generate a higher order bit more randomly than a lower order bit.

7.3.3 River Formation Dynamics

River formation dynamics (RFD) is a heuristic method similar to ant colony optimization. In fact, RFD can be seen as a gradient version of ACO, based on copying how water forms rivers by eroding the ground and depositing sediments. As water transforms the environment, altitudes of places are dynamically modified, and decreasing gradients are constructed. The gradients are followed by subsequent drops to create new gradients, reinforcing the best ones. By doing so, good solutions are given in the form of decreasing altitudes. This method has been applied to solve different NP-complete problems, such as the problems of finding a minimum distances tree and finding a minimum spanning tree in a variable-cost graph.

In fact, RFD fits particularly well for problems consisting of forming a kind of covering tree.

The basic outline of the RFD algorithm follows:

```
initializeDrops()

initializeNodes()

while   (not   allDropsFollowTheSamePath())   and   (not
otherEndingCondition())

   moveDrops()

   erodePaths()

   depositSediments()

   analyzePaths()

end while
```

The outline illustrates the key ideas of the algorithm. Firstly, all drops are put in the initial node. Then all nodes of the graph are initialized. This comprises mainly two acts: the altitude of the destination node is fixed to 0, and the altitude of the remaining nodes is set to some equal value. The `while` loop of the algorithm is executed until either all drops traverse the same sequence of nodes, or another alternative finishing condition is satisfied. The first step in the loop moves the drops across the nodes of the graph in a partially random way. In the next step, paths are eroded according to the movements of drops in the previous step. If a drop moves from a node P to a node Q, then we erode P. To be precise, the altitude of this node is reduced depending on the current gradient between P and Q. The erosion process prevents drops following cycles, as a cycle must include at least one slope and drops cannot climb up them. Once the erosion process finishes, the altitude of all nodes of the graph is somewhat augmented. Ultimately, the last step analyses all solutions found by drops and deposits the best solution found to date.

7.3.4 Particle Swarm Optimization

The particle swarm concept was motivated by the simulation of social behaviour. PSO requires only primitive mathematical operators, and is computationally inexpensive in terms of both memory requirements and time. The particles exhibit fast convergence to local and/or global optimal position(s) over a small number of generations.

A swarm in PSO consists of a number of particles. Each particle represents a potential solution to the optimization task. Each particle moves to a new position according to the new velocity, which includes its previous velocity, and the moving vectors according to the past best solution and global best solution. The best solution is then kept; each particle accelerates in the directions of not only the local best solution, but also the global best position. If a particle discovers a new probable solution, other particles will move closer to it in order to explore the region. Let's denote the swarm size. In general, there are three attributes, the particles' current position, current velocity, and past best position, for particles in the search space to present their features. Each particle in the swarm is updated according to the aforementioned attributes. The principle of the adaptive PSO is described as follows.

Figure 7.2 illustrates the flowchart of a PSO algorithm. During the PSO process, each potential solution is represented as a particle with a position vector x, referred to as phase weighting factor b and a moving velocity represented as v, respectively. So, for an K dimensional optimization, the position and velocity of the i th particle can be represented as $\mathbf{b}_i = (b_{i,1}, b_{i,2}, \ldots, b_{i,K})$ and $\mathbf{V}_i = (v_{i,1}, v_{i,2}, \ldots, v_{i,K})$, respectively. Each particle has its own best position $b_i^p = (b_{i,1}, b_{i,2}, \ldots, b_{i,K})$ corresponding to the individual best objective value obtained so far at time t,

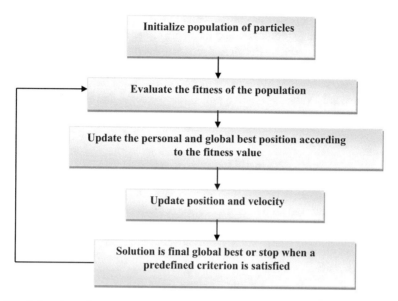

Fig. 7.2 A flowchart of the PSO algorithm

referred to as *p best*. The global best (*g best*) particle is denoted by $\mathbf{b}^G = \left(b_{g,1}, b_{g,2}, \ldots, b_{g,K}\right)$, which represents the best particle four[2] thus far at time t in the entire swarm. The new velocity $\mathbf{v}_i(t+1)$ for particle I is updated by

$$v_i(t+1) = wv_i(t) + c_1r_1\left(b_i^p(t) - b_i(t)\right) + c_2r_2\left(b^G(t) - b_i(t)\right),$$

where w is called *inertia weight*, $v_i(t)$ is the old velocity of the particle i at time t. Obviously, from this equation, the new velocity is related to the old velocity weighted by w and also associated to the position of the particle itself and that of the global best particle by acceleration constants c_i and c_2. The acceleration constants c_i and c_2 in the above equation adjust the amount of tension in PSO system. Low values allow particles to roam far from target regions before being tugged back, while high values result in abrupt movement toward, or past, target regions. The acceleration constants c_i and c_2 are therefore referred to as the cognitive and social rates, respectively, because they represent the weighting of the acceleration terms that pull the individual particle toward the personal best and global best positions. The velocities of the particles are confined in $[v_{min}, v_{max}]$. If an element of the velocity exceeds the thresholds v_{min} and v_{max}, it is set to the corresponding threshold.

The *inertia weight w* is employed to manipulate the impact of the previous history of velocities on the current velocity. A large inertia weight facilitates searching a new area, while a small inertia weight facilitates fine-searching in the current search area. Suitable selection of the inertia weight provides a balance between global exploration and local exploitation, and results in fewer iterations on average to find a sufficiently good solution. For the purpose of intending to simulate the slightly unpredictable component of natural swarm behaviour, two random functions r_1 and r_2 are applied to independently provide uniform distributed numbers in the range [0, 1] to stochastically vary the relative pull of the personal and global best particles. Based on the updated velocities, new position for particle i is computed according the following equation:

$$b_i(t + 1) = b_i(t) + v_i(t + 1)$$

The populations of particles are then moved according to the new velocities and locations calculated, and tend to cluster together from different directions. Thus, the evaluation of associated fitness of the new population of particles begins again. The algorithm runs through these processes iteratively until it stops.

Real World Case 2: Swarm Intelligence in Traffic Systems
Traffic problems are arising in several cities and handling these problems is a very complex issue, because the traffic information changes in real time and all transportation systems are affected by the number of vehicles, weather conditions, accidents, and so on. Therefore, there is a need to develop intelligent transport that can handle traffic efficiently and safely. Swarm intelligence is an effective way to manage these traffic-related data, and to obtain functional information.

7.3.5 Stochastic Diffusion Search

Stochastic diffusion search (SDS) is an agent-based probabilistic global search and optimization technique best suited to problems where the objective function can be decomposed into multiple independent partial-functions. It relies on several concurrent partial evaluations of candidate solutions by a population of agents and communication between those agents to locate the optimal match to a target pattern in a search space.

In SDS, each agent maintains a hypothesis that is iteratively tested by evaluating a randomly selected partial objective function parameterized by the agent's current hypothesis. In the standard version of SDS, such partial function evaluations are binary, resulting in each agent becoming active or inactive. Information on hypotheses is diffused across the population via inter-agent communication.

A basic outline of the SDS algorithm follows:

```
Initialisation: All agents generate an initial hypothe-
sis

while Halting criteria not satisfied do

  Test: All agents perform hypothesis evaluation

  Diffusion: All agents deploy a communication strategy

  Relate (Optional): active agents with the same hypothe-
  sis randomly deactivate

  Halt: Evaluation of halting criteria

end while
```

As a first step, agents' hypothesis parameters need to be initialized. Different initialization methods exist, but their specification is not needed for basic understanding of the algorithm.

All agents evaluate their hypothesis by randomly selecting one or a few micro-features from the target, mapping them into the search space using the transformation parameters defined by their hypothesis, and comparing them with the corresponding micro-features from the search space. Based on the result of the comparison, agents are divided into two groups: active or inactive. Active agents have successfully located one or more micro-features from the target in the search space; inactive agents have not.

In the diffusion step, each inactive agent picks at random another agent for communication. If the selected agent is active, then the selecting agent replicates its hypothesis. If the selected agent is inactive, then there is no flow of information between agents; otherwise, the selecting agent adopts a new random hypothesis. On the other hand, active agents do not start a communication session in standard SDS.

Unlike the communication used in ACO, in SDS, agents communicate hypotheses via a one-to-one communication strategy analogous to the tandem running procedure observed in some species of ant. A positive feedback mechanism ensures that, over time, a population of agents stabilizes around the global best solution.

7.3.6   Swarm Intelligence and Big Data

Swarm intelligence has become an obvious new field of nature-inspired computational intelligence. It takes its motivation from the collective behaviour of large numbers of individuals who collaborate with one another and their environment to achieve a common target.

Big data analytics is required to manage immense amounts of data quickly. The analytic problem can be modelled as optimization problems (Cheng et al. 2013). In general, optimization is concerned with finding the 'best available' solution(s) for a given problem within allowable time, and the problem may have several or numerous optimum solutions, of which many are local optimal solutions.

There are four major potentials of swarm intelligence:

1. *Optimization*: The selection of a best solution (with regard to some criteria) from some set of available alternatives. For example, particle swarm optimization can be employed in the engineering design problem of urban water resources planning.
2. *Clustering*: Gathering a set of objects together based on their similarities. For example, the application of an artificial bee colony for group creation and task allocation in rescue robots.
3. *Scheduling*: Allocating shifts and breaks to organization staff, based on their own desires and restrictions. For example, particle swarm optimization is used to solve staffing problems in large workplaces such as hospitals and airports.
4. *Routing*: Discovering the best set of paths for traversing from source to destination. Ant colony optimization is commonly used for vehicle routing problems.

In a nutshell, with swarms we can include more intelligence into the data-using tokens with a combination of intelligence and information. Swarm intelligence is a stimulating area of research and several companies are using these techniques.

Real World Case 3: Energy Management using Swarm Intelligence

Internet of Things (IoT) and swarm logic are expected to play a vital role in efficient energy management in the future. With the rising cost of commercial energy, smart energy management systems will be used by organizations for competitive advantage. As the amount of data collected by smart meters increases over time, these systems will increasingly leverage the latent information hidden in the data to make smart choices. Big data and metaheuristics approaches are expected to be widely used in the process as data scientists begin to mine such data to develop smarter algorithms and decision making tools. Robust swarm algorithms will be applied across varied scenarios, to identify and prioritize energy optimization opportunities, and to coordinate activities without any single device issuing orders. For instance, the devices communicate wirelessly and use swarming algorithms to collaboratively decide how to manage power usage, i.e. reducing energy consumption by mimicking the self-organizing behavior of bees.

7.4   Case-Based Reasoning

Case-based reasoning (CBR) is different from usual predictive analytics. Where predictive analytics uses statistical methods to be able to determine potential outcomes or potential risks to an organization, CBR does not actually rely on so many statistics.

Case-based reasoning is a problem-solving paradigm that in many respects is fundamentally different from even major AI approaches (Aamodt and Plaza 1994). Instead of relying solely on general knowledge of a problem domain, or making associations along generalized relationships between problem descriptors and conclusions, CBR is able to utilize the specific knowledge of previously experienced, concrete problem situations (cases). A new problem is solved by finding a similar past case, and reusing it in the new problem situation. A second important difference is that CBR is also an approach to incremental, sustained learning, since a new experience is retained each time a problem has been solved, making it immediately available for future problems.

Case-based reasoning (CBR) tries to solve new problems by reusing specific past experiences stored in example cases. A case models a past experience, storing both the problem description and the solution applied in that context. All the cases are stored in the case base. When the system is presented with a new problem to solve, it searches for the most similar case(s) in the case base and reuses an adapted version of the retrieved solution to solve the new problem.

To be specific, CBR is a cyclic and integrated problem solving process (see Fig. 7.3) that supports learning from experience and has four main steps: retrieve, reuse, adaptation and retain (Kolodner 1993). The adaptation phase is split into two substeps: revise and review. In the revise step, the system adapts the solution to fit the specific constraint of the new problem. In the review step, the constructed solution is evaluated by applying it to the new problem, understanding where it fails and making the necessary corrections.

Sometime the solution retrieved can be straightforwardly reused in the new problem, but in the majority of situations, the retrieved solution is not directly applicable and must be adapted to the specific requirements of the new problem. After this adaptation, the system creates a new case and can retain it in the case base (learning).

> **Real World Case 4: Case Based Reasoning in Oil & Gas Industry**
> Drilling an offshore well usually involves huge investments and high daily costs. The drilling industry is an extremely technology-oriented industry. Hence, any kinds of tools or equipment that can improve the drilling operation

(continued)

are crucial in all stages of operation. Case-based reasoning has been shown to provide effective support for several tasks related to drilling. Optimization of drilling plans, which are highly repetitive, can be achieved through CBR. The CBR method is also used for solving operational problems. Information gathered via the problem analysis process may be used in order to make decisions on how to proceed further. Cases in a CBR system can be stored to predict an upcoming situation through evolving inconsistent sequences of measurement data.

Moreover, in the oil exploration field, CBR is beneficial in running pattern matching over real-time data streams from operating equipment. Data is matched against earlier known problems and the associated risks of each are computed. Decision makers can monitor risk levels and take action whenever necessary to keep risks within controllable limits. Once a risk threshold is crossed, alerts are triggered and operators are notified that urgent action is required.

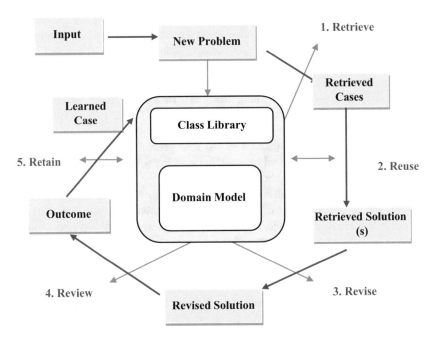

Fig. 7.3 Case-based reasoning cycle

7.4.1 Learning in Case-Based Reasoning

If reasoning (and learning) is directed from memory, then learning answers to a process of prediction of the conditions of case recall (or retrieval). Learning in CBR answers a nature of the system to anticipate future situations: the memory is directed toward the future both to avoid situations having caused a problem and to reinforce the performance in success situations (Akerkar 2005).

7.4.1.1 Case Representation

A case is a contextualized piece of knowledge representing an experience that teaches a lesson fundamental to archiving the goal of the reasoner.

A central issue in CBR is the case model. This must account for both the problem and solution components. It is necessary to decide which attributes should compose a case and what representation language is better suited to represent the particular knowledge involved in the problem solving process. Hence, the case representation task is concerned with:

1. The selection of pertinent attributes,
2. The definition of indexes, and
3. Structuring the knowledge in a specific case implementation.

Another way to describe case presentations is to visualize the structure in terms of the problem space and the solution space. According to this structure, the description of a problem resides in the problem space. The retrieval process identifies the features of the case with the most similar problem. When best matching is found, the system uses similarity metrics to find the best matching case. In those processes, the solution of a case with the most similar problem may have to be adapted to solve the new problem.

7.4.1.2 Case Indexing

Indexing is related to the creation of additional data structures that can be held in the memory to speed up the search process, focussing on the most relevant dimensions. The indexes identify the case attributes that should be used to measure case similarity. Moreover, indexes can speed up the retrieval process by providing fast access to those cases that must be compared with the input case problem.

Methodologies for choosing indexing include manual and automated methods. In some systems, cases are indexed by hand. For instance, when the cases are complex and the knowledge needed to understand cases well enough to choose indexes accurately is not concretely available, hand indexing is needed. On the other hand, if problem solving and understanding are already automated, it is advantageous to use automated indexing methods.

7.4.1.3   Case Retrieval

Case retrieval is a process in which a retrieval algorithm retrieves the most similar cases to the current problem. Case retrieval requires a combination of search and matching. In general, two retrieval techniques are used by the major CBR applications: nearest neighbour retrieval algorithm and inductive retrieval algorithm.

Nearest-Neighbour Retrieval

Nearest-neighbour retrieval algorithm is a commonly used similarity metric in CBR that computes the similarity between stored cases and new input case based on weight features. A typical evaluation function is used to compute nearest-neighbour matching as:

$$\text{similarity} (\text{Case}_I, \text{Case}_R) = \frac{\sum_{i=1}^{n} w_i \times sim \left(f_i^I, f_i^R \right)}{\sum_{i=1}^{n} w_i}$$

Where w_i is the importance weight of a feature, sim is the similarity function of features, and f_i^I and f_i^R are the values for feature i in the input and retrieved cases, respectively. A major weakness of the algorithm is its efficiency. This is predominantly weak for large case bases or those of high dimensionality.

Inductive Retrieval

Inductive retrieval algorithm is a technique that determines which features do the best job in discriminating cases, and generates a decision tree type structure to organize the cases in memory. This approach is very useful when a single case feature is required as a solution, and when that case feature is dependent upon others.

Nearest neighbour retrieval and inductive retrieval are widely applied in CBR applications and tools. The choice between nearest neighbour retrieval and inductive retrieval in CBR applications requires experience and testing. Applying nearest neighbour retrieval is appropriate without any pre-indexing; however, when retrieval time turns out to be a key issue, inductive retrieval is preferable. Nearest neighbour retrievals have many uses in addition to being a part of nearest neighbour classification. For example, when biologists identify a new protein, they use a computer program to identify, in a large database of known proteins, the proteins that are the most similar to the new protein.

7.4.2 Case-Based Reasoning and Data Science

Predictive analytics platforms enable organizations to leverage all enterprise data—from historical structured data to the latest unstructured big data—to drive faster, more informed decision making and provide preventive warnings of systems failure. Here, CBR as a part of predictive analytics uses a learning approach to solve current problems with knowledge gained from past experience. A CBR-driven predictive analytics engine obtains patterns by automatically and continuously comparing real-time data streams of multiple heterogeneous data types. To proactively direct the user to the most appropriate decision or action, a (self-learning) case library adapts past solutions to help solve a current problem and recognizes patterns in data that are similar to past occurrences.

Using CBR, systems can learn from the past and become more adaptive. For example, if a system begins exhibiting a pattern of inconsistent trading behaviour, CBR searches for past cases of similar patterns and issues an alert for action before the pattern intensifies into a risky consequence.

There are several successful applications of CBR in the financial service sector. The most significant opportunity for CBR is to serve as an early alert system for market operators to prevent disruptions caused by outages, trading errors, or other compliance violations. Capital markets institutions can deploy CBR to monitor and deter abnormal client behaviour, detect risk exposures through internal or external illegal activities, improve IT operational efficiency in the back office and uncover customer-facing opportunities to make profits.

7.4.3 Dealing with Complex Domains

Uncertainty in CBR can occur due to three main reasons. First, information may simply be missing. For example, the problem domain may be so complex that it can only be represented incompletely. Even in simpler domains, it is not always appropriate to describe a complex situation in every detail, but to tend to use the functionality and the ease of acquisition of the information represented in the case as criterion to decide the representation of a case. Second, for different problems, different features of the world and the problem description will play different roles in achieving a solution. Third, perfect prediction is impossible. There is no immediate solution to remove/reduce this kind of uncertainty from the problem. The best we can do is to select a course of action according to our expectation and understanding about the current situation, then keep track of the state of the world, learn more about the situation if possible, and dynamically adjust the actions.

Finally, the shortcomings of CBR can be summarised as follows:

- *Handling large case bases*: High storage requirements and time-consuming retrieval accompany CBR systems utilizing large case bases.
- *Dynamic problem domains*: CBR systems may have difficulties in dealing with dynamic problem domains, where they may be unable to follow a shift in the way problems are solved, since they are most strongly biased towards what has already been performed.
- *Handling noisy data*: Parts of the problem situation may be irrelevant to the problem itself. In turn, this indicates inefficient storage and retrieval of cases.
- *Fully automatic operation*: In a usual CBR system, the problem domain is typically not fully covered. Therefore, some problem situations can occur for which the system has no solution.

7.5 Rough Sets

Rough set theory is a useful means for studying delivery patterns, rules, and knowledge in data. The rough set is the estimate of a vague concept by a pair of specific concepts, called the lower and upper approximations (Akerkar and Lingras 2007). The classification model signifies our knowledge about the domain.

Let us assume that our set of interest is the set S, and we understand which sample elements are in S. We would like to define S in terms of the attributes. The membership of objects with respect to a random subset of the domain may not be definable. This fact gives rise to the definition of a set in terms of lower and upper approximations. The *lower approximation* is a type of the domain objects which are known with certainty to belong to the subset of interest. The *upper approximation* is a description of the objects which may perhaps belong to the subset. Any subset defined through its lower and upper approximations is called a *rough set*, if the boundary region is not empty. Now we present the formal definition of this concept.

Suppose an equivalence relation, θ on U, is a binary relation that is transitive, reflexive and symmetric. In rough set theory, an equivalence relation is known an *indiscernibility relation*. The pair (U, θ) is called an approximation space. With each equivalence relation θ, there is a partition of U such that two elements x, y in U are in the same class in this partition, if and only if $x\theta y$. Let us represent a class in the partition due to θ, as $\theta_x = \left\{ y \in U \middle| x\theta y \right\}$. For a subset $X \subseteq U$, we say that

- $\underline{X} = \cup \left\{ \theta_x \middle| \theta_x \subseteq X \right\}$ is said to be the lower approximation or positive region of X.
- $\overline{X} = \cup \left\{ \theta_x \middle| x \in X \right\}$ is said to be the upper approximation or possible region of X.
- The *rough set* of X is the pair $\left(\underline{X}, \overline{X} \right)$.
- $\left(\underline{X} - \overline{X} \right)$ is the area of uncertainty.
- $\underline{X} \cup \left(U - \overline{X} \right)$ is said to be the region of certainty.

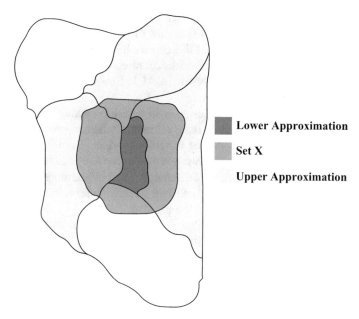

Fig. 7.4 Rough sets

Figure 7.4 illustrates the concept of rough set theory.

We can describe rough set theory using a database T, which is a set of tuples. The set of attributes on these tuples is defined as $A = \{A_1, A_2, \ldots, A_m\}$. For a tuple $t \in T$ and for a subset $X \subseteq A$, $t[X]$ denotes the projection of the tuple t, on the set of attributes in X.

For a given subset of attributes Q, we define an equivalence relation on T as follows. For two tuples t_1, t_2, we say that $t_1 \theta t_2$ if $t_1[Q] = t_2[Q]$. Specifically, we say that two tuples t_1 and t_2 are imperceptible with respect to the attributes in Q. Furthermore, we can determine the lower and upper approximation for any subset of tuples in T. We can say that a set of attributes Q is dependent on another set P, if the partition of the database with respect to P contains the partition with respect to Q. This will lead to efficient techniques of attribute elimination required for decision trees, association rules and clustering.

Let us assume that the database T contains the following transactions: $\{a, b, c\}, \{a, b, d\}, \{a, c, d\}, \{a, c, e\}, \{a, d, e\},$ and $\{a, d, f\}$. Let us further assume that the items in the transaction are ordered.

We define an equivalence relation between transactions as having two common prefixes. In other words, two transactions are equivalent if their first two elements are the same. Now let us define the lower and upper approximations for X, which contains transactions $\{a, b, c\}, \{a, b, d\},$ and $\{a, c, d\}$. The lower approximation of X will be given by $\{\{a, b, c\}, \{a, b, d\}\}$. The upper approximation of X will be $\{\{a, b, c\}, \{a, b, d\}, \{a, c, d\}, \{a, c, e\}\}$.

The notion of the rough set was proposed in 1982 by Zdzislaw Pawlak, (Pawlak 1982), and has been found useful in the realms of knowledge acquisition and data mining. Many practical applications of this approach have been developed in recent years, in areas such as medicine, pharmaceutical research, and process control. One of the primary applications of rough sets in AI is for the purpose of knowledge analysis and discovery of data.

One may note that managing uncertainty in decision-making and prediction is very crucial for mining any kind of data, no matter whether small or big. While fuzzy logic is well known for modelling uncertainty arising from vague, ill-defined or overlapping concepts/regions, rough sets model uncertainty due to granularity (or limited discernibility) in the domain of discourse. Their effectiveness, both individually and in combination, has been established worldwide for mining audio, video, image and text patterns, as present in the big data generated by different sectors. Fuzzy sets and rough sets can be further coupled, if required, with (probabilistic) uncertainty arising from randomness in occurrence of events, in order to result in a much stronger framework for handling real life ambiguous applications. In the case of big data, the problem becomes more acute because of the manifold characteristics such as high varieties, velocity, variability and incompleteness.

7.6   Exercises

1. Describe the term recency with respect to tabu search.
2. Describe how ants are able to find the shortest path to a food source.
3. Briefly describe the main principles of a case-based reasoning, its operation process and possible differences in implementation.
4. Describe characteristics of problems in which it is better to use the case-based approach.
5. Name and briefly describe three main sources of uncertainty.
6. Consider the problem of finding the shortest route through several cities, such that each city is visited only once and in the end one returns to the starting city. Suppose that in order to solve this problem we use a genetic algorithm, in which genes represent links between pairs of cities. For example, a link between Oslo and Madrid is represented by a single gene 'LP'. Let also assume that the direction in which we travel is not important, so that $LP = PL$.

   (a) How many genes will be used in a chromosome of each individual if the number of cities is 10?
   (b) How many genes will be in the alphabet of the algorithm?

7. (*Project*) Explore how metaheuristics can be used as an optimization technique to precisely analyse large data in the industry sector of your choice.
8. (*Project*) Metaheuristic approaches have focused on finding high quality solutions without performance guarantee in a reasonable amount of time. Investigate different metaheuristic approaches for big data clustering.

References

Aamodt, A., & Plaza, E. (1994). Case-based reasoning: Foundational issues, methodological variations and system approaches. *AI Communications, 17*(1), 39–59.

Akerkar, R. (2005). *Introduction to artificial intelligence.* PHI Learning.

Akerkar, R., & Lingras, P. (2007). *Building an intelligent web: Theory & practice.* Sudbury: Jones & Bartlett Publisher.

Cheng, S., Yuhui, S., Quande, Q., & Ruibin, B. (2013). *Swarm intelligence in big data analytics.* s.l. (Lecture notes in computer science, pp. 417–426). Berlin/Heidelberg: Springer.

Gendreau, M. (2002). *Recent advances in Tabu search. I: Essays and surveys in metaheuristics* (pp. 369–377). s.l.: Kluwer Academic Publishers.

Glover, F., & Laguna, M. (1997). *Tabu search.* Norwell: Kluwer Academic Publishers.

Hertz, A., & de Werra, D. (1991). The Tabu search metaheuristic: How we used it. *Annals of Mathematics and Artificial Intelligence, 1*, 111–121.

Kennedy, J., & Eberhart, R. (2001). *Swarm intelligence.* London: Academic.

Kolodner, J. (1993). *Case-based reasoning.* San Francisco: Morgan Kaufmann.

Laguna, M., & Mart'I, R. (2003). *Scatter search – Methodology and implementations in C.* Norwell: Kluwer Academic Publishers.

Pawlak, Z. (1982). Rough sets. *International Journal of Parallel Programming, 11*(5), 341–356.

Taillard, E., Gambardella, L., Gendreau, M., & Potvin, J. (2001). Adaptive memory programming: A unified view of metaheuristics. *European Journal of Operational Research, 135*, 1–16.

Teodorovic´, D., & Dell'Orco, M. (2005). *Bee colony optimization – A cooperative learning approach to complex transportation problems,* Poznan: 10th EWGT Meeting.

Tereshko, V., & Loengarov, A. (2005). Collective decision-making in honeybee foraging dynamics. *Computing and Information Systems Journal, 9*(3), 1–7.

Chapter 8
Analytics and Big Data

8.1 Introduction

In this chapter, we outline some advanced tools and technologies, including the Apache Hadoop ecosystem, real-time data streams, scaling up machine learning algorithms, and fundamental issues such as data privacy and security. We cover much of the basic theory underlying the field of big data analytics in this chapter, but of course, we have only scratched the surface. Keep in mind that to apply the concepts contained in this overview of big data analytics, a much deeper understanding of the topics discussed herein is necessary.

Data nowadays stream from daily life: from phones and credit cards and televisions and telescopes; from the infrastructure of cities; from sensor-equipped buildings, trains, buses, planes, bridges, and factories. These trends are popularly referred to as *big data*.

Analysis of this big data to derive knowledge, therefore, requires data-driven computing, where the data drives the computation and control, including complex queries, analysis, statistics, intelligent computing, hypothesis formulation and validation.

In 2001, Gartner[1] introduced the 3 Vs definition of data growth that was then in the inception stages:

- *Volume*: The volume of data can be extremely high in the event of years of data collection. The data volume build-up may be from unstructured sources like social media and machine-based data that can be collected.
- *Variety*: Data collection can be from varied sources and forms. It can be sourced out from emails, audio and video forms.

---

[1]http://www.gartner.com/technology/home.jsp

© Springer International Publishing Switzerland 2016
R. Akerkar, P.S. Sajja, *Intelligent Techniques for Data Science*,
DOI 10.1007/978-3-319-29206-9_8

- *Velocity*: The velocity of data streaming is quite fast paced. Organizations are often overwhelmed in embracing the loads of information that are available to them, and managing them is a big challenge in today's digital era.

Recently, business and industry acknowledge 5 Vs, adding Veracity and Value as additional aspects of data.

- *Veracity*: The messiness or trustworthiness of the data. With many forms of big data, quality and accuracy are less controllable, but big data and analytics technology now allows us to work with this type of data.
- *Value*: The data received in original form generally has a low value relative to its volume. However, a high value can be obtained by analysing large volumes of such data.

Big data has given rise to associated problems that cut across all phases of the data lifecycle.

Volume and Scalability This is the fundamental problem that every tool tackles when dealing with big data. Therefore, big data tools and infrastructure need to ensure adequate scalability and flexibility to be able to handle the super speed of data growth.

Heterogeneous and Unstructured Data Most big data is unstructured and therefore heterogeneous in nature. Analytic tools therefore need to be smart enough to interpret the diverse natures of data, integrate them with advanced algorithm development, and optimize to bring them in a consistent, consumable format.

Data Governance and Security Industries such as banking, healthcare, telecommunication and defence are under strict compliance and regulatory mandates, making it difficult to create an appropriate data protection framework. Data governance has taken key prominence in many industry sectors where opportunity is enormous in big data, but risks can be huge.

Infrastructure and System Architecture As the cutting-edge technologies of Hadoop and MapReduce are scaled to meet the 5 Vs of big data, they assert significant demands on infrastructure in terms of scale and storage capacities that are efficient and cost effective.

However, though a few innovators such as Google, Facebook, and Amazon were utilizing big data with success, big data was largely unexplored terrain for the majority of mainstream businesses. For these companies, it is the ability to integrate more *sources* of data than ever before—small data, big data, smart data, structured data, unstructured data, social media data, behavioural data, and legacy data. This is the *variety* challenge, and has appeared as the highest data priority for mainstream companies. Thus, tapping into more data sources has emerged as the latest data frontier within the business world.

8.2 Traditional Versus Big Data Analytics

Traditional analytics typically involves extension and novel integration of existing analytical approaches, rather than the creation of new analytical methods. Successful data analysis solutions are driven by businesses' need for insightful and actionable intelligence and focus on producing relevant and accurate evidence informing such action.

Traditional data analysis, therefore, uses proper statistical and machine learning methods to analyse huge data, to concentrate, extract, and refine useful data hidden in a batch of chaotic data sets, and to identify the inherent law of the subject matter, so as to maximize the value of data.

> **Real World Case 1: Retail Customer Profiling**
> Retail customer profiling is where e-commerce stores, operators, and warehouse managers want to find immediate insights on their customer's changing needs, interests and behaviour, so they can customize their services and offer promotional or upsell items at the precise point in time to the right customer.

Some representative traditional data analysis techniques are discussed in Chap. 3. In the following list, we add a few more traditional techniques.

- *Classification*: This analysis identifies to which of a prescribed set of classes a newly observed data point belongs. The classes may be explicitly defined by users, or may be learned by a classification algorithm based on other features of the data.
- *Cluster Analysis*: A statistical method for grouping objects, and specifically, classifying objects according to certain features. Cluster analysis is used to differentiate objects with particular features and divide them into some categories (clusters) according to these features, such that objects in the same category will have high homogeneity, while different categories will have high heterogeneity. As we have discussed earlier, cluster analysis is an unsupervised study method without training data. In a nutshell, clustering is similar to classification, but labels data points based on their similarity (in one or more features) with each other. Data points are then grouped into likely classes based on metrics that may be deterministic (e.g., k-means), statistical (e.g., multi-variate distributions and likelihood maximization), or network graph-based (e.g., centrality and betweenness).
- *Regression Analysis*: A mathematical tool for revealing correlations between one variable and several other variables. Based on a group of experiments or observed data, regression analysis identifies dependence relationships among variables hidden by randomness. Regression analysis may make complex and undetermined correlations among variables to be simple and regular.

- *Factor Analysis*: This analysis is basically targeted at describing the relation among many elements with only a few factors, i.e., grouping several closely related variables into a factor, and the few factors are then used to reveal most of the information of the original data.
- *Correlation Analysis*: An analytical method for determining the law of relations, such as correlation, correlative dependence, and mutual restriction, among observed phenomena and accordingly conducting forecast and control.
- *Bucket Testing*: A method for determining how to improve target variables by comparing the tested group. Big data will require a large number of tests to be executed and analysed.
- *Statistical Analysis*: Statistical analysis is based on statistical theory, where randomness and uncertainty are modelled with probability theory. Statistical analysis can provide a description and an inference for big data. Descriptive statistical analysis can summarize and describe data sets, while inferential statistical analysis can draw conclusions from data subject to random variations. Statistical analysis is widely applied in the economic and medical care fields.
- *Predictive Modelling*: Predictive modelling not only attempts to forecast future events, it also seeks to identify ways to optimize future outcomes. The underlying characteristics of the original data determine the types of meaningful predictions and optimizations that are possible to generate.
- *Pattern Recognition and Matching*: Pattern matching seeks to identify trends, sequences, and patterns in data sets. The patterns may be explicitly defined in advance, learned directly from data, and/or manually refined after initial learning from data.
- *Information Retrieval*: This technique uses direct manual or automated queries of a data set to answer the question: What in this data set is relevant to the query? It typically relies on matching the query to explicitly represented features of the data without inference or prediction based on hidden features.

Big data analytics can be deemed as the technique for a special kind of data. Thus, many traditional data analysis methods may still be utilized for big data analysis. However, to extract key insights from big data, we need advanced big data processing methods, as listed below.

- *Parallel Computing*: This refers to concurrently employing several computing resources to complete a computation task. Its basic idea is to decompose a problem, which is then assigned to several separate processes to be independently completed, so as to achieve co-processing. There are some classic parallel computing models developed, including MPI (Message Passing Interface), MapReduce, and Dryad.
- *Hashing*: This method essentially transforms data into shorter fixed-length numerical values or index values. A hash is like a fingerprint for data. A hash function takes your data as an input, and gives you back an identifier of a smaller, fixed length, which you can use to index or compare or identify the data. Advantages of hashing are rapid reading, writing, and high query speed; however, it is challenging to find a sound Hash function.

- *Bloom Filter*: A bloom filter is a simple space-efficient randomized data structure for representing a set in order to support membership queries. A bloom filter consists of a series of hash functions, a data structure that tests the containment of a given element in a set. In other words, a bloom filter stores hash values of data other than data itself, by utilizing a bit array. It has advantages such as high space efficiency and high query speed, but, at the same time, has some shortcomings in misrecognition and deletion.
- *Index*: An index is an effective method to reduce the overhead of disk reading and writing, and to improve insertion, deletion, modification, and query speeds in relational databases that manage structured data, as well as in other technologies that manage semi-structured and unstructured data. However, an index has shortcomings in that it has the extra cost of storing index files that should be dynamically maintained when data is updated.
- *Trie tree*: This is a variant of Hash Tree. It is mainly applied to rapid retrieval and word frequency statistics. The main idea of trie is to utilize common prefixes of character strings to reduce comparison on character strings to the greatest extent, so as to improve query efficiency. We use a trie to store pieces of data that have a *key* (used to identify the data) and possibly a *value* (which holds any additional data associated with the key).

8.3 Large-Scale Parallel Processing

With *parallel programming*, we break up the processing workload into multiple parts, which can be executed concurrently on multiple processors. Not all problems can be parallelized. The challenge is to identify as many tasks that can run concurrently as possible. Alternatively, we can identify data groups that can be processed concurrently. This will allow us to divide the data among multiple concurrent tasks.

One example of the kind of application that is worth targeting is the MapReduce programming framework. The MapReduce model allows developers to write massively parallel applications without much effort, and is becoming an essential tool in the software stack of many companies that need to deal with large data sets. MapReduce fits well with the idea of dynamic provisioning, as it may run on a large number of machines and is already widely used in cloud environments.

8.3.1 MapReduce

MapReduce is a programming model used to develop massively parallel applications that process and generate large amounts of data. It was first introduced by Google in 2004 (Dean and Ghemawat 2004), and has since become an important tool for distributed computing. It is suited to operate on large data sets on clusters of computers, as it is designed to tolerate machine failures.

A MapReduce program is composed of three steps:

- `Map()` step, in which the master node imports input data, parses them in small subsets, and distributes the work on slaves nodes. Any slave node will produce the intermediate result of a `map()` function, in the form of key/value pairs, which are saved on a distributed file. Output file location is notified to the master at the end of the mapping phase.
- `Shuffle` step, in which the master node collects the answers from slave nodes, to combine the key/value pairs in value lists sharing the same key, and sort them by the key. Sorting can be lexicographical, increasing or user- defined.
- `Reduce()` step performs a summary operation.

That means, MapReduce divides the work into small computations in two major steps, `map` and `reduce`, which are inspired by similar primitives that can be found in functional programming languages. A logical view of MapReduce is shown in Fig. 8.1.

The user specifies a `map` function that processes a key/value pair to generate a set of intermediate key/value pairs, and a `reduce` function that merges all intermediate values associated with the same intermediate key.

The `map` function emits each word plus an associated count of occurrences. That function is then applied to each of the values. For example:

$$\left(\texttt{map}'\ \texttt{length}'\left(()\ (a)\ (abc)\ (abcd)\right)\right)$$

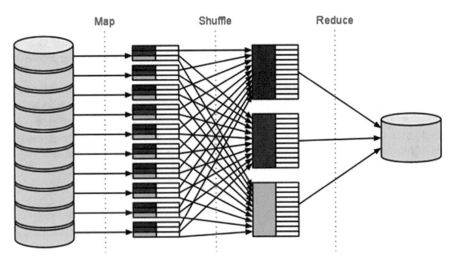

Fig. 8.1 MapReduce job

applies the length function to each of the three items in the list. Since *length* returns the length of an item, the result of map is a list containing the length of each item:

$$(0\ 1\ 34)$$

The reduce function is given a binary function and a set of values as parameters. It combines all the values together using the binary function. If we use the + (add) function to reduce the list (0 1 34):

$$\left(\text{reduce }\#'+'(0\ 1\ 34)\right)$$

we get

$$8$$

Here each application of the function to a value can be performed in parallel, since there is no dependence of one upon another. The reduce operation can take place only after the map is complete.

MapReduce has facilitated standardizing parallel applications. It is also powerful enough to solve a broad variety of real world problems, from web indexing to pattern analysis to clustering.

During the map phase, nodes read and apply the map function to a subset of the input data. The map's partial output is stored locally on each node, and served to worker nodes executing the reduce function.

Input and output files are usually stored in a distributed file system, but in order to guarantee scalability, the master tries to assign local work, meaning the input data is available locally. Conversely, if a worker node fails to carry out the work it has been assigned to complete, the master node is able to send the work to some other node.

MapReduce is implemented in a master/worker configuration, with one master serving as the coordinator of many workers. A worker may be assigned a role of either map *worker* or reduce *worker*.

1. *Split input*: The input splits can be processed in parallel by different machines.
2. *Fork processes*: The next step is to create the master and the workers. The master is responsible for dispatching jobs to workers, keeping track of progress, and returning results. The master chooses idle workers and assigns them either a map task or a reduce task. A map task works on a single shard of the original data. A reduce task works on intermediate data generated by the map tasks.
3. *Map*: Each map task reads from the input shard that is assigned to it. It parses the data and generates *(key, value)* pairs for data of interest. In parsing the input, the map function is likely to get rid of a lot of data that is of no interest. By having many map workers do this in parallel, we can linearly scale the performance of the task of extracting data.

4. *Map worker – Partition*: The stream of *(key, value)* pairs that each worker generates is buffered in memory and periodically stored on the local disk of the map worker. This data is partitioned into R regions by a partitioning function.
5. *Reduce Sort*: When all the map workers have completed their work, the master notifies the reduce workers to start working. At this point a reduce worker gets the data that it desires to present to the user's reduce function. The reduce worker contacts every map worker via remote procedure calls to get the *(key, value)* data that was targeted for its partition. When a reduce worker has read all intermediate data, it sorts it by the intermediate keys so that all occurrences of the same key are grouped together. The sorting is needed because typically many different keys map to the same reduce task. If the amount of intermediate data is too large to fit in memory, an external sort is used.
6. *Reduce function*: With data sorted by keys, the user's Reduce function can now be called. The reduce worker calls the Reduce function once for each unique key. The function is passed two parameters: the key and the list of intermediate values that are associated with the key.
7. *End*: When all map tasks and reduce tasks have been completed, the master wakes up the user program. At this moment, the MapReduce call in the user program returns back to the user code. Output of MapReduce is stored.

8.3.2 Comparison with RDBMS

Relational Database Management Systems (RDBMS) are the dominant choice for transactional and analytical applications, and they have traditionally been a well-balanced and sufficient solution for most applications. Yet its design has some limitations that make it difficult to keep the compatibility and provide optimized solutions when some aspects such as scalability are the top priority.

There is only a partial overlap of functionality between RDBMSs and MapReduce: relational databases are suited to do things for which MapReduce will never be the optimal solution, and vice versa. For instance, MapReduce tends to involve processing most of the data set, or at least a large part of it, while RDBMS queries may be more fine-grained. On the other hand, MapReduce works fine with semi-structured data since it is interpreted while it is being processed, unlike RDBMSs, where well-structured and normalized data are the key to ensuring integrity and improve performance. Finally, traditional RDBMSs are more suitable for interactive access, but MapReduce is able to scale linearly and handle larger data sets. If the data are large enough, doubling the size of the cluster will also make running jobs twice as fast, something that is not necessarily true of relational databases.

Another factor that is also driving the move toward other kind of storage solutions are disks. Improvements in hard drives seem to be relegated to capacity and transfer rate only. But data access in a RDBMS is usually dominated by seek times, which haven't changed significantly for some years.

MapReduce has been criticized by some RDBMS supporters, due to its low-level abstraction and lack of structure. But taking into account the different features and goals of relational databases and MapReduce, they can be seen as complementary rather than opposite models.

8.3.3 Shared-Memory Parallel Programming

Traditionally, many large-scale parallel applications have been programmed in shared-memory environments such as OpenMP.[2] OpenMP offers a set of compiler directives to create threads, synchronize the operations, and manage the shared memory on top of `pthreads`. The programs using OpenMP are compiled into multi-threaded programs, in which threads share the same memory address space, and hence the communications between threads can be very efficient.

Compared to using `pthreads` and working with `mutex` and condition variables, OpenMP is much simpler to use because the compiler takes care of transforming the sequential code into parallel code according to the directives. Therefore, programmers can write multi-threaded programs without deep understanding of multi-threading procedure.

Compared to MapReduce, these kinds of programming interface are much more generic and provide solutions for a wider variety of problems. One of the mainstream use cases of these systems is for parallel applications that require some kind of synchronization.

The main difference between MapReduce and this model is the hardware for which each of these platforms has been designed. MapReduce is supposed to work on commodity hardware, while interfaces such as OpenMP are only efficient in shared-memory multiprocessor platforms.

8.3.4 Apache Hadoop Ecosystem

Hadoop is a popular and widely used open source MapReduce implementation. It has a large community base and is also backed and used by companies such as Yahoo!, IBM, Amazon, and Facebook. Hadoop was originally developed by Doug Cutting to support distribution for the Nutch search engine. The first working version was available by the end of 2005, and soon after that, in early 2006, Doug Cutting joined Yahoo! to work on it full-time with a dedicated team of developers. In February 2008, Yahoo! announced that they were using a 10,000-core Hadoop cluster in production to generate their search index.

---

[2]http://openmp.org/wp/

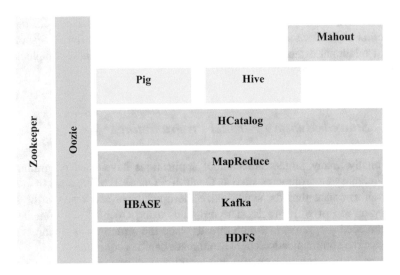

Fig. 8.2 Hadoop ecosystem

In April 2008, Hadoop was able to sort a terabyte of data on a 910-node cluster in 209 s. That same year in November, Google managed to break that record by a wide margin with a time of 68 s on a 1000-node cluster. Hadoop is now a top-level Apache project, and hosts a number of subprojects including HDFS, Pig, HBase and ZooKeeper. The Hadoop ecosystem is illustrated in Fig. 8.2.

Since its first releases, Hadoop[3] has been the standard free software MapReduce implementation. Even though there are other open source MapReduce implementations, they are not as complete as some component of the full platform (e.g., a storage solution). Hadoop is currently a top-level project of the Apache Software Foundation, a non-profit corporation that supports a number of other well-known projects such as the Apache HTTP Server (White 2012).

Hadoop is mostly known for its MapReduce implementation, which is in fact a Hadoop subproject, but there are also other subprojects that provide the required infrastructure or additional components. MapReduce software provides framework for distributed processing of large data sets on compute clusters of commodity hardware.

A brief description of the different components of Hadoop ecosystem follows (Akerkar 2013):

- HDFS: Hadoop Distributed File System – storage, replication.
- MapReduce: Distributed processing, fault tolerance.

---

[3]Hadoop is licensed under the Apache License 2.0, a free-software license that allows developers to modify the code and redistribute it.

- HBase: Fast read/write access – a column-oriented database management system that runs on top of HDFS.
- HCatalog: Metadata – a table and storage management layer for Hadoop that enables users with different data processing tools.
- Pig: Scripting – The salient property of Pig programs is that their structure is amenable to substantial parallelization, which in turns enables them to handle very large data sets.
- Hive: SQL – A data warehouse infrastructure built on top of Hadoop for providing data summarization, query, and analysis.
- Oozie: Workflow, scheduling to manage Apache Hadoop jobs.
- Zookeeper: Coordination – a centralized service for maintaining configuration information, naming, providing distributed synchronization, and providing group services.
- Kafka: Cluster retains all published messages—whether or not they have been consumed—for a configurable period of time.
- Mahout: Produces implementations of distributed or otherwise scalable machine learning.

Real World Case 2: Optimizing Marketing Websites
Clickstream data is an important part of big data marketing. It tells retailers what customers click on and do or do not purchase. However, storage for viewing and analysing these insights on other databases is expensive, or they just do not have the capacity for all of the data exhaust. Apache Hadoop is able to store all the web logs and data for years, and at low cost, allowing retailers to understand user paths, conduct basket analysis, run bucket tests, and prioritize site updates, thus improving customer conversion and revenue.

HDFS is a distributed file system that runs on large clusters and provides high throughput access to application data. The remaining subprojects are simply additional components that are usually used on p of the core subprojects to provide additional features. Some of the most noteworthy are:

Pig is a high-level data flow language and execution framework for parallel computation. Programs written in this high-level language are translated into sequences of MapReduce programs. This sits on top of Hadoop and makes it possible to create complex jobs to process large volumes of data quickly and efficiently. It supports many relational features, making it easy to join, group, and aggregate data.

HBase is a column-oriented database management system that runs on top of HDFS and supports MapReduce computation. It is well suited for sparse data sets. HBase does not support a structured query language like SQL. An HBase system comprises a set of tables. Each table contains rows and columns, much like a traditional database. Each table must have an element defined as a Primary Key, and all access attempts to HBase tables must use this Primary Key. An HBase column represents an attribute of an object.

Hive is data warehouse infrastructure that provides data summarization and ad-hoc querying and analysis of large files. It uses a language similar to SQL, which is automatically converted to MapReduce jobs. Hive looks like conventional database code with SQL access. Though Hive is based on Hadoop and MapReduce operations, there are many differences. Hadoop is aimed for long sequential scans, and because Hive is based on Hadoop, you can anticipate queries to have a high latency. This means that Hive would not be appropriate for applications that need very fast response times, as you would expect with a database such as DB2. Moreover, Hive is read-based and not suitable for transaction processing that normally involves a high percentage of write operations.

Chukwa is a Hadoop subproject devoted to large-scale log collection and analysis (Boulon et al. 2008). Chukwa is built on top of the Hadoop distributed file system (HDFS) and MapReduce framework, and inherits Hadoop's scalability and robustness. It is a data collection and monitoring system for managing large distributed systems; it stores system metrics as well as log files into HDFS, and uses MapReduce to generate reports.

8.3.5 Hadoop Distributed File System

The Apache Hadoop Common library is written in Java and consists of two main components: the MapReduce framework and Hadoop Distributed File System (HDFS[4]), which implements a single writer, multiple reader model. Nevertheless, Hadoop does not merely support HDFS as an underlying file system. The goal of HDFS is to store large data sets reliably and to stream them at high bandwidth to user applications. HDFS has two types of nodes in the schema of a master–worker pattern: a `namenode`, the master, and an arbitrary number of `datanodes`, the workers. The HDFS namespace is a hierarchy of files and directories with associated metadata represented on the `namenode`. The actual file content is fragmented into blocks of 64 MB, where each block is typically duplicated on three `namenodes`. The `namenode` keeps track of the namespace tree and the mapping of file blocks to `datanodes`. An HDFS client who wishes to read a file has to contact the `namenode` for the locations of data blocks and then reads the blocks from the neighbouring `datanode`, since HDFS considers short distance between nodes as higher bandwidth between them. In order to keep track of the distances between `datanodes`, HDFS supports rack–awareness. Once a `datanode` registers with the `namenode`, the `namenode` runs a user–configured script to decide which network switch (rack) the node belongs to. Rack–awareness also allows HDFS to have a block placement policy that provides a trade-off between minimizing write

---

[4]http://hadoop.apache.org/hdfs/

cost and maximizing data reliability, availability and aggregate read bandwidth. For the formation of a new block, HDFS places the first replica on the `datanode` hosting the writer, and the second and third replicas on two different `datanodes` located in a different rack. A Hadoop MapReduce job, a unit of work that the client wants to be performed, consists of the input data (located on the HDFS), the MapReduce program and configuration information.

Built-in Hadoop MapReduce programs are written in Java; however, Hadoop also provides the Hadoop streaming application programming interface (API), which allows writing `map` and `reduce` functions in languages other than Java by using Unix standard streams as the interface between Hadoop and the user program.

In Hadoop, there are two types of nodes that control the job execution process: one job tracker and an arbitrary number of task trackers. The job tracker coordinates a job run on the system by dividing it into smaller tasks to run on different task trackers, which in turn transmit reports to the job tracker. In case a task fails, the job tracker is able to automatically reschedule the task on a different available task tracker. In order to have a task tracker run a map task, the input data needs to be split into fixed-size pieces. Hadoop runs one map task for each split, with the user-defined map function processing each record in the split. As soon as a map task is accomplished, its intermediary output is written to the local disk. After that, the map output of each map task is processed by the user-defined reduce function on the reducer. The number of map tasks running in parallel on one node is user-configurable and heavily dependent on the capability of the machine itself, whereas the number of reduce tasks is specified independently and is therefore not regulated by the size of the input. In case there are multiple reducers, one partition per reducer is created from the map output. Depending on the task to accomplish, it is also possible to have zero reduce tasks in case no reduction is desired.

8.4   NoSQL

NoSQL, known as 'Not only SQL database', provides a mechanism for storage and retrieval of data and is the next generation database. NoSQL states various benefits associated with its usage.

Before going into the details of NoSQL,[5] let us recall our understanding about traditional databases. In the relational model, there are three basic components defining a data model:

1. *Data structure*: the data types a model is built of.
2. *Operators*: the available operations to retrieve or manipulate data from this structure.
3. *Integrity rules*: the general rules defining consistent states of the data model.

---

[5]http://nosql-database.org/

The structure of a relational data model is mainly given by relations, attributes, tuples and (primary) keys. Relations are usually visualized as tables, with attributes as columns and tuples as rows. The order of the attributes and tuples is not defined by the structure, thus it can be arbitrary. Basic operations defined by the relational model are SELECT operations (including projections and joins) to retrieve data, as well as manipulative operations such as INSERT, UPDATE and DELETE. Two different sets of integrity rules can be distinguished for a relational model. Constraints such as uniqueness of primary keys ensure the integrity within a single relation. Furthermore, there are referential integrity rules between different relations.

A notable idea in relational database systems is transactions. There are basically three properties of a transaction: atomicity, consistency and durability. Later, Harder and Reuter abbreviated those properties together with a new one: isolation – by the acronym ACID. Even though all four ACID properties are seen as key properties of transactions on relational databases, consistency is particularly interesting when investigating the scalability of a system. The scalability of a system is its potential to cope with an increasing workload.

Principally, a system can be scaled in two different directions: vertically and horizontally. Vertical scaling ('scale up') means increasing the capacity of the system's nodes. This can be achieved by using more powerful hardware. In contrast, a system is scaled horizontally ('scaled out') if more nodes are added.

An important difference between relational databases and NoSQL databases is in the provided consistency models. NoSQL systems have to soften the ACID guarantees given by relational transactions, in order to allow horizontal scalability.

In NoSQL, there are four prominent data models:

- Key-Value
- Document
- Column-Family
- Graph

They are schema-less, i.e., no need to define a precise schema before start storing data, as you need with relational databases. Moreover, the first three share a common characteristic, i.e., they use keys to store *aggregates* as values. Here, aggregate is a collection of related objects. The databases that support aggregates (i.e., complex records) are very straightforward and fast to work with, because you do not need to normalize your data as you typically do on relational databases.

8.4.1   Key-Value Model

The key-value model is a kind of hash table, where keys are mapped to values. To use this model, you principally use a unique key (a kind of ID, usually a string) to store an aggregate value. Aggregates on key-value databases are opaque. This means we can store several data types, but can only retrieve through a key without query support. Key-value databases are the simplest NoSQL solutions. Due to their simplistic nature and usually great performance, they are heavily used not as a primary storage, but rather as some kind of cache for highly accessed data. Other common use cases include job queue, real-time analysis and session management.

Examples are Memcached, Redis, Riak, and Vedis.

8.4.1.1 Document Model

The document model uses key-value to store data, but the database can see the structure of the aggregate. It pairs each key with a complex data structure known as a document. It can contain many different key-value pairs or key-array pairs or even nested documents. Document databases have broader use cases than key-value ones. They can be used on content management systems and content-based websites, user generated content, monitoring systems, product catalogue, online gaming and social-network apps.

Examples are CouchDB, MongoDB, RethinkDB.

8.4.1.2 Column-Family Model

The column-family model has characteristics of the key-value, document and relational model. Column-family databases were designed to handle massive amounts of data; thus, they run on clusters (distributed), providing high availability with no single point of failure.

Examples are Bigtable, Cassandra, Hbase, and SimpleDB.

8.4.1.3 Graph Model

The graph model is a totally different model than the others. It was built to satisfy a scenario where there are many relationships between objects. They are used to store information about networks, such as social connections. The graph model treats objects as nodes and relationships as edges, and both can have attributes. Graph databases usually do not run on clusters, but instead on only one machine, as relational databases.

Graph model examples are Neo4J and HyperGraphDB.

In a nutshell, the relational data model, with its complex set of operators and integrity rules, offers a very flexible storage solution, which can be adapted to many problems. Moreover, relational transactions, having ACID properties, give strong assurances on consistency. On the other hand, NoSQL databases aim at delivering easy and efficient solutions for particular problems.

8.5 SPARK

Apache Spark[6] is an open source, big data processing framework built around speed, ease of use, and sophisticated analytics. It was originally developed in 2009 in UC Berkeley's AMPLab, and open sourced in 2010 as an Apache project. Spark's machine learning library, called MLlib,[7] provides capabilities that are not easily utilized in Hadoop MapReduce without the use of a general-purpose computation engine like Spark. These machine learning algorithms are able to execute faster as they are executed in-memory, as opposed to MapReduce programs, which have to move data in and out of the disks between different stages of the processing pipeline.

Spark is a framework that provides a highly flexible and general-purpose way of dealing with big data processing needs, does not impose a rigid computation model, and supports a variety of input types (Hamstra and Zaharia 2013). This enables Spark to deal with text files, graph data, database queries, and streaming sources, and not be confined to a two-stage processing model. Programmers can develop arbitrarily complex, multi-step data pipelines arranged in an arbitrary directed acyclic graph (DAG) pattern.

Spark has several advantages compared to other big data and MapReduce technologies such as Hadoop and Storm. Programming in Spark involves defining a sequence of *transformations* and *actions*. Spark has support for a *map* action and a *reduce* operation, so it can implement traditional MapReduce operations, but it also supports SQL queries, graph processing, and machine learning. Unlike MapReduce, Spark stores its intermediate results in memory, providing for dramatically higher performance.

Spark takes MapReduce to the next level with less expensive shuffles in the data processing. With capabilities such as in-memory data storage and near real-time processing, the performance can be several times faster than other big data technologies. Spark also supports lazy evaluation of big data queries, which helps with optimization of the steps in data processing workflows. It provides a higher level API to improve developer productivity and a consistent architect model for big data solutions.

Spark holds intermediate results in memory rather than writing them to disk, which is very useful, especially when you need to work on the same data set multiple times. It's designed to be an execution engine that works both in-memory and on-disk. Spark operators perform external operations when data does not fit in memory. Spark can be used for processing data sets that are larger than the aggregate memory in a cluster.

Spark architecture (see Fig. 8.3) includes the following three main components:

Data Storage Spark uses HDFS file system for data storage purposes. It works with any Hadoop compatible data source including HDFS, HBase and Cassandra.

---

[6]https://spark.apache.org/

[7]http://spark.apache.org/mllib/

Fig. 8.3 Spark architecture

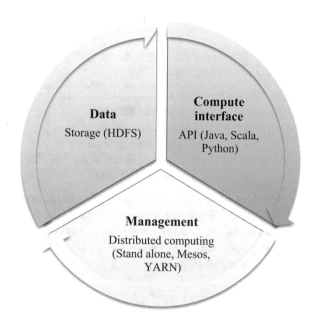

API The API provides the application developers to create Spark based applications using a standard API interface. Spark provides API for Scala, Java, and Python programming languages.

Resource Management Spark can be deployed as a Stand-alone server or it can be on a distributed computing framework such as Mesos or YARN.

Spark is a complementary technology that should be used in partnership with Hadoop. Remember that Spark can run separately from the Hadoop framework, where it can integrate with other storage platforms and cluster managers. It can also run directly on top of Hadoop, where it can easily leverage its storage and cluster manager.

Spark features include:

- Supports more than just Map and Reduce functions.
- Optimizes arbitrary operator graphs.
- Lazy evaluation of big data queries, which helps with the optimization of the overall data processing workflow.
- Provides concise and consistent APIs in Scala, Java and Python.
- Offers interactive shell for Scala and Python. This is not available in Java yet.

Spark is written in Scala[8] Programming Language and runs on Java Virtual Machine (JVM) environment.

---

[8]http://www.scala-lang.org/

8.6   Data in Motion

There are significant challenges to analysing big data and making it actionable. Some of the challenges are:

- The size (volume) of data.
- The variety of data types and sources.
- The need for fast and yet complex analyses.
- The velocity of new data being generated.
- The need for flexibility of operations; for instance, the ability to quickly create or modify views or filters on this data.

Looking from a different perspective, the industry examples also bring to light that they have a big data problem with the 'data at rest'—the historical data they have had in their repositories, and 'data in motion'—the data they are generating on a non-stop basis from their operations and interactions. Remarkably, both categories are growing persistently.

> **Real World Case 3: Location-Based Mobile Wallet Services**
> In location-based mobile wallet service, a telco service provider can offer its customers location-specific information guided by historical preferences regarding cuisine, entertainment and shopping, in real time. For instance, sending coupons for favourite products just as the customer nears a grocery store.

8.6.1   Data Stream Processing

Data stream processing is the *in-memory* (i.e., a database management system that primarily relies on main memory for computer data storage), record-by-record analysis of machine data in motion. The objective is to extract actionable intelligence as streaming analytics, and to react to operational exceptions through real-time alerts and automated actions in order to correct or avert the problem.

> **Real World Case 4: Insurance Data Streams**
> An insurance company needs to compare the patterns of traffic accidents across a wide geographic area with weather data. In these cases, the analysis has to be fast and practical. Also, companies analyse the data to see whether new patterns emerge.

Streaming data is useful when analytics need to be done in real time while the data is in motion. In fact, the value of the analysis (and often the data) decreases with time. For example, if you cannot analyse and act immediately, a sales opportunity might be lost or a threat might go undetected.

Several industry sectors are benefiting from stream processing; for example:

- *Smart Cities*: real-time traffic analytics, congestion prediction and travel time apps.
- *Oil & Gas*: real-time analytics and automated actions to avert potential equipment failures.
- *Telecommunication*: real-time call rating, fraud detection and network performance data.
- *Security intelligence*: for fraud detection and cybersecurity alerts such as detecting Smart Grid consumption issues, SIM card misuse and SCADA intrusion attacks.
- *Industrial automation*: offering real-time analytics and predictive actions for patterns of manufacturing plant issues and quality problems.

The high-level architecture of a stream processor is closely related to the types of dataflow architectures exhibited by signal processing hardware systems, where arriving data streams are clocked through a network of processing nodes, and where each node performs some action or transformation on the data as it flows through. Software stream processing behaves in exactly the same way, except instead of logic gates, each node in the system is a continuously executing and independent query that performs operations on data streams such as filtering, aggregation and analytics.

The conceptual architecture for a stream processor is a network of concurrently executing continuous queries that operate on the data streams as they flow through.

8.6.2 Real-Time Data Streams

Batch processing systems (e.g., Hadoop) had developed for offline data processing platform for big data. As we have discussed, Hadoop is a high-throughput system that can crunch a huge volume of data using a distributed parallel processing paradigm called MapReduce. But there are many situations across diverse domains that require real-time response on big data for speedy decision making (Ellis 2014). For example, credit card fraud analytics need to process real time data stream on the fly to predict if a given transaction is a fraud. Hadoop is not appropriate for those situations. If decisions such as this are not taken in real time, the opportunity to mitigate the damage is lost. Therefore, there is a need for real-time processing systems that perform analytics on short time windows, i.e., correlating and predicting events streams generated for the last few minutes. In order to have enhanced prediction capabilities, real-time systems often leverage batch-processing systems such as Hadoop.

Real World Case 5: Security in Thermal Power Station
A thermal power station wants to be an extremely secure environment so that unauthorized persons do not intervene with the delivery of power to clients. Power stations place sensors around the perimeter of a site to detect movement. In this case, the enormous amount of data coming from these sensors needs to be analysed in real time so that an alarm is sounded only when a genuine threat exists.

In concert with the volume and variety, real-time processing also needs to handle the velocity of the data. There are three steps in this regard:

1. The real-time processing system should be able to collect the data generated by real-time events streams coming in at a rate of millions of events per seconds.
2. It needs to handle the parallel processing of this data as and when it is being collected.
3. It should perform event correlation using a complex event processing engine to extract the meaningful information from this moving stream.

These three steps should happen in a fault tolerant and distributed way. The real-time processing system should be a low latency system so that the computation can happen speedily, with almost real-time response capabilities.

8.6.3   Data Streams and DBMS

By now you know that a data stream is a continuous sequence of items produced in real-time. A stream can be considered to be a relational table of infinite size. It is therefore considered impossible to maintain an order of the items in the stream with respect to an arbitrary attribute. Likewise, it is impossible to store the entire stream in memory. However, results of operations are expected to be produced as soon as possible.

Real World Case 6: Gaming Data Feed
Real time data processing tools can be used to continuously collect data about player–game interactions and feed the data into your gaming platform. With such tools, one can design a game that provides engaging and dynamic experiences based on players' actions and behaviours.

As a consequence, standard relational query processing cannot be straightforwardly applied, and online stream processing has become its own field of research in the area of data management. Common examples where online stream processing is important are network traffic monitoring, sensor data, web log analysis, online auctions, inventory and supply-chain analysis and real-time data integration.

Conventional Database Management Systems (DBMSs) are designed using the concept of persistent and interrelated data sets. These DBMSs are stored in reliable repositories, which are updated and queried frequently. But there are some modern application domains where data is generated in the form of a stream and Data Stream Management Systems (DSMSs) are required to process the stream data continuously.

The basic difference between a conventional DBMS and a DSMS is the nature of query execution. In DBMSs, data is stored on disk and queries are performed over persistent data. In DSMS, in contrast, data items arrive online and stay in the memory for short intervals of time. DSMSs need to work in non-blocking mode while executing a sequence of operations over the data stream.

To accommodate the execution of a sequence of operations, DSMSs often use the concept of a window. A window is basically a snapshot taken at a certain point in time and it contains a finite set of data items. When there are multiple operators, each operator executes and stores its output in a buffer, which is further used as an input for some other operator. Each operator needs to manage the contents of the buffer before it is overwritten.

Common operations performed by most DSMSs are filtering, aggregation, enrichment, and information processing. A stream-based `join` is required to perform these operations. A stream-based `join` is an operation that combines information coming from multiple data sources. These sources may be in the form of streams or they may be disk-based. Stream-based `joins` are important components in modern system architectures, where just-in-time delivery of data is expected. One example is an online auction system that generates two streams, one stream for opening an auction, while the other stream consists of bids on that auction. A stream-based join can relate the bids with the corresponding opened auction in a single operation.

Real World Case 7: Customer Satisfaction in Telecommunication Industry

A telecommunications company in a very competitive market wants to make sure that outages are carefully monitored so that a detected drop in service levels can be escalated to the appropriate group. Communications systems generate huge volumes of data that have to be analysed in real time to take the appropriate action. A delay in detecting an error can seriously impact customer satisfaction.

8.7   Scaling Up Machine Learning Algorithms

Most of the machine learning algorithms belong to the mid-1970s, when the computational resources and the datasets size were limited, so those algorithms are often not scalable enough to handle big data. Scalability is indispensable when the data set stops fitting on a single machine and the processing becomes unacceptably slow. If you plan ahead for your processing needs, much hassle can be avoided. For example, all scalable solutions should be implemented for at least two machines. That way most of the scaling problems are solved in the initial implementation if more processing power is needed.

The main reasons for scaling up machine learning algorithms are:

- Large number of data instances: the number of training examples is extremely large.
- High input dimensionality: the number of features is very large, and may need to be partitioned across features.
- Model and algorithm complexity: a number of high accuracy algorithms are computationally expensive, and rely on either complex routines or non-linear models.
- Inference time constraints: applications such as robotic navigation require real time prediction.
- Model selection and parameter sweeps: Tuning hyper parameters of learning algorithms and statistical evaluation require multiple executions of learning and inference.

Real World Case 8: Aircraft Engine Condition Monitoring
Airline companies pay per hour of 'time on wing', a measure of operational reliability of the aircraft engine. This causes engine makers to enhance the engine's reliability. Here, machine learning methods perform pattern matching for fault isolation and repair support using multiple operational and external parameters received in real-time from sensor data. This facilitates engine makers in accurately predicting failure in engine operations well ahead of time, as a result intensifying the service revenue and reducing the cost of service. This enables engine industry to predict the health of the equipment in real time, allowing customers to release the equipment for maintenance only when necessary. Machine learning algorithms such as neural networks, support vector machines, and decision trees are efficient in identifying complex interdependencies within operational parameters and uncovering anomalies that can lead to equipment failures.

Machine learning algorithms are considered the core of data driven models. However, scalability is considered crucial requirement for machine learning algorithms as well as any computational model.

Two techniques basically are applicable to scale up machine learning algorithms.

First, the parallelization of the existing sequential algorithms is what Apache Mahout and Apache Spark follow to scale up the machine learning algorithms. Getting Mahout to scale effectively isn't as straightforward as simply adding more nodes to a Hadoop cluster. Factors such as algorithm choice, number of nodes, feature selection, and sparseness of data—as well as the usual suspects of memory, bandwidth, and processor speed—all play a role in determining how effectively Mahout can scale. Apache Mahout aims to provide scalable and commercial machine learning techniques for large-scale and intelligent data analysis applications. Many renowned companies, such as Google, Amazon, Yahoo!, IBM, Twitter, and Facebook have implemented scalable machine learning algorithms in their projects. Many of their projects involve big data problems and Apache Mahout provides a tool to alleviate the big challenges. Mahout's core algorithms, including clustering, classification, pattern mining, regression, dimension reduction, evolutionary algorithms and batch-based collaborative filtering, run on top of a Hadoop platform via the Map/reduce framework.

Real World Case 9: Recommending Fitness Products

A product recommender App collects fitness data for users through smart sensors, finds similar users based on the fitness data, and recommends fitness products to the users. This application caters to sensors from different manufacturers, and uses sensors such as a sleep sensor (keeps track of sleep quality), an activity sensor (tracks steps, distance, calories burnt, and stairs climbed), and Wi-Fi smart scale (tracks weight and synchronizes user statistics). For the recommendation engine, the application uses the user similarity model (helps to compare fitness trends of similar users). One crucial challenge is that the recommendation data set needs to be updated in real time. Here Mahout is suitable for heavy lifting. The interface offers a refresh strategy. When it is invoked, it takes care of refreshing all the components right down to the data model. The data model is implemented on MongoDB to scale the refresh capability. Mahout's user similarity model maintains a list of similar users for every user known to the recommendation engine. So, the application asks Mahout for similar users and plotting their fitness trends.

Second, redesign the structure of existing models to overcome the scalability limitation. The result of this technique is new models that extend the existing ones, such as the Bag-of-Words (CBOW) model. Google researchers proposed the CBOW model as an extension to the feed-forward Neural Network Language Model (NNLM) by removing the non-linear hidden layer, which caused most of the complexity of the original model. This extension allows the new model to handle big data efficiently, which the original model was not suitable for.

8.8   Privacy, Security and Ethics in Data Science

Security and privacy concerns are growing as data becomes more and more accessible. The collection and aggregation of massive quantities of heterogeneous data are now possible. Large-scale data sharing is becoming routine among scientists, clinicians, businesses, governmental agencies, and citizens. However, the tools and technologies that are being developed to manage these massive data sets are often not designed to integrate sufficient security or privacy measures, in part because we lack sufficient training and a fundamental understanding of how to provide large-scale data security and privacy. We also lack adequate policies to ensure compliance with current approaches to security and privacy. Furthermore, existing technological approaches to security and privacy are increasingly being breached, thus demanding the frequent checking and updating of current approaches to prevent data leakage. While the aggregation of such data presents a security concern in itself, another concern is that these rich databases are being shared with other entities, both private and public.

In many cases, potentially sensitive data are in the hands of private companies. For some of these companies such as Google, Facebook, and Instagram, the data about their users is one of their main assets, and already a part of the product that they sell (known as targeted marketing). But even if this is not the company's main business, the ability to protect the privacy and data of its customers and users will represent a major risk. Potential privacy issues, data protection, and privacy-related risks should therefore be addressed early in any big data project.

Data scientists use big data to find out about our shopping preferences, health status, sleep cycles, moving patterns, online consumption and friendships. In some cases, such information is individualized. Removing elements that allow data to be connected to one individual is, however, just one feature of anonymization. Location, gender, age, and other information pertinent to the belongingness to a group and thus valuable for data analytics relate to the issue of group privacy. To strip data from all elements concerning to any kind of group belongingness would mean to strip it from its content. Therefore, regardless of the data being anonymous in the sense of being not individualized, groups are often becoming more transparent.

In order to protect the user privacy, best practices in the prevention and detection of abuse by continuous monitoring must be implemented. Privacy-preserving analytics is an open area of research that can help minimize the success of malicious actors from the data set. However, there are few practical solutions at the moment.

> **Real World Case 10: Fraud and Compliance**
> Data can detect and deal with security issues before they become problems. In real life, security situations and compliance requirements are often evolving,

(continued)

and data can help reveal dubious activity and mitigate risk in a manner that was previously not possible. Analysing data can reduce the operational costs of fraud investigation, help anticipate and prevent fraud, streamline regulatory reporting and compliance. Doing so effectively requires aggregating and analysing data from multiple sources and types, and analysing it all at once —imagine financial transaction data, geo-location data from mobile devices, commercial data, and authorization and data. With the ability to analyse all data, including point of sale, geo-location and transaction data, credit card companies are able to spot potential fraud more accurately than ever before.

Differential privacy is a good first step toward privacy preservation. Differential privacy defines a formal model of privacy that can be implemented and proven secure at the cost of adding computational overhead and noisy results to data analytics results. Perhaps the current definition of differential privacy is too conservative, and a new, more practical definition might address some of costs associated with the implementation of this principle.

Despite privacy challenges, the utilization of big data could also have huge benefits for society at large. By using demographic and mobility data, we can obtain key insights into human behaviour, including traffic patterns, crime trends, crisis responses, and social unrest. These, in turn, can be used by business and policy makers to create better, safer, and more efficient societies.

8.9 Exercises

1. Describe the five Vs that characterize big data application. Further explore utilization of big data technologies within a data-driven industry sector; for example, Twitter, Facebook, eBay or Disney.
2. Distinguish the different types of NoSQL databases.
3. Explore strengths and limitations of Hadoop, Pig, Hive, Chukwa, and HBase.
4. Identify three real world use cases for Hadoop.
5. Compare and contrast relational database management systems (RDBMS) and HBase.
6. What is the impact of the cluster on database design?
7. There are many ways retailers interact with customers, including social media, store newsletters and in-store. But customer behaviour is almost completely unpredictable without Hadoop. Justify.
8. Search the news for an article about privacy breaches. Briefly describe the reported incident and discuss its privacy implications.
9. (*Project*) A physician can only access medical records of his patient. The physician is authorized to access medical records during her duty at the hospital.

Accesses are authorized only if the patient has given his informed consent. A physician can access patient records without patient consent only in the case of emergency. How will you as a data scientist devise a policy supporting this scenario?

10. (*Project*) Select various NoSQL and Hadoop products and conduct a comparative study (also include merits and limitations) of the tools that you have selected.

11. (*Project*) Prepare a position article (or essay) to highlight the privacy challenges associated with big data, primarily within the realm of the smart cities sector. The article should be intended for decision makers, public authorities, industry and civil society.

References

White, T. (2012). *Hadoop: The definitive guide.* s.l.: Yahoo Press.

Akerkar, R. (2013). *Big data computing.* s.l.: Chapman and Hall/CRC.

Boulon, J. et al. (2008). *Chukwa, a large-scale monitoring system* (pp. 1–5). Chicago: s.n.

Dean, J., & Ghemawat, S. (2004). *MapReduce: Simplified data processing on large clusters* (pp. 137–150). San Francisco, CA, s.n.,.

Ellis, B. (2014). *Real-time analytics: Techniques to analyze and visualize streaming data.* s.l.: Wiley.

Hamstra, M., & Zaharia, M. (2013). *Learning spark.* s.l.: O'Reilly Media.

Chapter 9
Data Science Using R

As we have seen in Chap. 1 and Appendices, there are several free and commercial tools available for exploring, modelling, and presenting data. Due to its usefulness and flexibility, the open source programming language R is used to demonstrate many of the analytical techniques discussed in this book. Familiarity with software such as R allows users to visualize data, run statistical tests, and apply machine learning algorithms. R runs on almost any standard computing platform and operating system. Its open source nature means that anyone is free to adapt the software to whatever platform they choose. Indeed, R has been reported to be running on modern tablets, phones, and game consoles. A key advantage that R has over many other statistical packages is its sophisticated graphics capabilities.

Although R is persuasive and flexible, with a powerful set of statistical algorithms and graphical capabilities, it is single threaded and in-memory. Therefore, it is hard to scale to large data sizes. Due to this shortcoming, data scientists mostly rely on sampling the big data sets sitting on a platform, and then perform analytics on the reduced data. Obviously, they do not gain valuable insights from the reduced data. One can integrate R's statistical capabilities with Hadoop's distributed clusters in two ways: interfacing with SQL query languages, and integration with Hadoop Streaming.

This chapter offers a concise overview of the fundamental functionality of the R language.

9.1 Getting Started

The primary R system is available from the Comprehensive R Archive Network,[1] also known as CRAN. CRAN also hosts many add-on packages that can be used to extend the functionality of R.

---

[1]https://cran.r-project.org/

© Springer International Publishing Switzerland 2016
R. Akerkar, P.S. Sajja, *Intelligent Techniques for Data Science*,
DOI 10.1007/978-3-319-29206-9_9

Once you have downloaded and installed R, open R to figure out your current directory, type `getwd()`. To change directory, use `setwd` (note that the 'C:' notation is for Windows and would be different on a Mac):

```
> setwd("C:\\Datasets")
```

The UC Irvine Machine Learning Repository[2] contains a couple hundred data sets, mostly from a variety of real applications in science and business. These data sets are used by machine learning researchers to develop and compare algorithms.

Moreover, you can easily extend R's basic capabilities by adding in functions that others have created and made freely available through the web. Rather than surfing the web trying to find, them, R makes adding these capabilities trivially easy by using the 'Packages' menu items.

For example, to add a new package not previously downloaded onto your computer, click on 'Packages → Install Package(s)', pick a site from the list and choose the package you want to install from the long list. To activate this package, you then need to click on it in the 'Packages → Load Package' menu item. You are then ready to use it, and the help files for the package will also be automatically added to your installation. See, for example, the 'Help → HTML help' menu item, and click on 'Packages' in the browser window that opens.

Another important thing is to determine the working directory that R uses to load and save files. You will need to select the working directory at the start of the session. This can be done with text commands, but is more simply accomplished through the menu bar. On a PC, click on change `dir` under the File option. On a Mac, click on Change Working Directory under the Misc option. You can also load text files directly from websites using read.csv, only with the website name in parentheses instead.

9.2 Running Code

You could use R by simply typing at the command prompt; however, this does not easily allow you to save, repeat, or share your code. So, go to 'File' in the top menu and click on 'New script'. This opens up a new window that you can save as an .R file. To execute the code, you type into this window, highlight the lines you wish to run, and press Ctrl-R on a PC or Command-Enter on a Mac. If you want to run an entire script, make sure the script window is on top of all others, go to 'Edit', and click 'Run all'. Any lines that are run appear in red at the command prompt.

When you test any data science algorithm, you should use a variety of datasets. R conveniently comes with its own datasets, and you can view a list of their names by typing `data()` at the command prompt.

---

[2]http://archive.ics.uci.edu/ml/datasets.html

9.3   R Basics

You will type R commands into the R console in order to carry out analyses in R. In the R console you will see:

>

This is the R prompt. We type the commands needed for a specific task after this prompt. The command is carried out after you press the Return key. Once you have started R, you can start typing in commands, and the results will be calculated immediately; for example:

```
> 2*3
[1]  6
> 10-3
[1]  7
```

All variables (scalars, vectors, matrices, etc.) created by R are called *objects*. R stores data in a couple different types of variables. The primary one is a *vector*, which is just a string of numbers. A *matrix* variable is a two-dimensional arrangement of numbers in rows and columns. Less common are *factor* variables, which are strings of text items (for example, names). If you have spreadsheets with your research data, they would most likely be considered a *data frame* in R, which is a combination of *vector* and *factor* variables arranged in columns. For example, you might have a column with sample codes (a factor variable), a column with measurement localities or rock types (a factor variable), and columns with measured numerical data (vector variables).

Some useful functions for getting information about data in variables include the following.

To see the data contained in a variable, simply type its name at the command line:

```
> variablename
```

To see the column names, without displaying data, in a data frame variable:

```
> names(dataframename)
```

In order to manually rename the columns and/or rows:

```
> colnames(dataframename)<-
       c("Column1","Column2","ColumnN")
```

```
> rownames(dataframename)<-c("Row1","Row2","RowN")
```

To see a list of the categories in a factor variable, arranged in alphabetic order:

```
> levels(factorname)
```

To count the number of items in a vector variable:

```
> length(vectorname)
```

To count the number of rows in a data frame:

```
> nrow(dataframename)
```

Furthermore, in R, we assign values to variables using an arrow. For example, we can assign the value 2*3 to the variable x using the command:

```
> x <- 2*3
```

To view the contents of any R object, just type its name, and the contents of that R object will be displayed:

```
> x
[1] 6
```

There many different types of objects in R, including scalars, vectors, matrices, arrays, data frames, tables, and lists. The scalar variable x is one example of an R object. While a scalar variable such as x has just one element, a vector consists of several elements. The elements in a vector are all of the same type (e.g., numeric or characters), whereas lists may include elements such as characters as well as numeric quantities.

To create a vector, we can use the c() (combine) function. For example, to create a vector called myvector that has elements with values 9, 6, 11, 10, and 4, we type:

```
> myvector <- c(9, 6, 11, 10, 4)
```

To see the contents of the variable myvector, you can type its name:

```
> myvector
[1]   8   6   9 10   5
```

The [1] is the index of the first element in the vector. We can extract any element of the vector by typing the vector name with the index of that element given in square brackets. For example, to get the value of the third element in the vector myvector, we type:

```
> myvector[3]
[1] 11
```

In contrast to a vector, a list can contain elements of different types; for example, both numeric and character elements. A list can also include other variables such as a vector. The list() function is used to create a list. For example, we can create a list mylist by typing:

```
> mylist <- list(name="James", wife="Rita",
  myvector)
```

We can then print out the contents of the list mylist by typing its name:

```
> mylist
$name
[1] "James"

$wife
[1] "Rita"

[[3]]
[1]   8   6   9 10   5
```

The elements in a list are numbered, and can be referred to using indices. We can extract an element of a list by typing the list name with the index of the element given in double square brackets. Hence, we can extract the second and third elements from mylist by typing:

```
> mylist[[2]]
[1] "Rita"
> mylist[[3]]
[1]   9, 6, 11, 10, 4
```

Elements of lists may also be named, and in this case the elements may be referred to by giving the list name, followed by "$", followed by the element name. For example, mylist$name is the same as mylist[[1]] and mylist$wife is the same as *mylist[[2]]*:

```
> mylist$wife
[1] "Rita"
```

We can find out the names of the named elements in a list by using the attributes() function, for example:

```
> attributes(mylist)
$names
[1] "name" "wife" ""
```

When you use the attributes() function to find the named elements of a list variable, the named elements are always listed under the heading '$names'. Therefore, we see that the named elements of the list variable mylist are called 'name' and 'wife', and we can retrieve their values by typing mylist$name and mylist$wife, respectively.

9.4 Analysing Data

R has simple built-in functions for calculating descriptive statistics about a sample, in order to estimate the central tendency (mean or median) and amount of dispersion in the data (variance or standard deviation). To compute the mean, variance, standard deviation, minimum, maximum, and sum of a set of numbers, use mean, var, sd, min, max, and sum. There are also rowSum and colSum to find the row and column sums for a matrix. To find the component-wise absolute value and square root of a set of numbers, use abs and sqrt. Correlation and covariance for two vectors are computed with cor and cov, respectively.

The problem most commonly encountered is that of missing data (empty cells in a spreadsheet, which are treated as NA by R). If you use a descriptive statistics function when there are NA values, the result will also be NA. However, all descriptive statistics have a built-in method to deal with NA values, so for mean(), median(), var(), or sd() you can add na.rm=TRUE in the command.

For example:

```
mean(x,na.rm=TRUE) #computes the mean while deleting NA
values

sd(x,na.rm=T) #you can abbreviate any criteria as long
as   it   is   a   unique   identifier   (T   is a unique
abbreviation of TRUE because it can't be confused for
FALSE or F).
```

As in other programming languages, you can write if statements, and for and while loops. For instance, here is a simple loop that prints out even numbers between 1 and 10 (%% is the modulo operation):

```
> for (i in 1:10){
    + if(i%%2==0){
    + cat(paste(i, "is even.\n", sep=" ")) #
use paste to concatenate strings
    +   }
    +}
```

The 1:10 part of the for loop can be specified as a vector. For instance, if you wanted to loop over indices 1, 2, 3, 5, 6, and 7, you could type for (i in c(1:3,5:7)). To pick out the indices of elements in a vector that satisfy a certain property, use 'which'; for example:

```
> which(v >= 0)      # indices of nonnegative
elements of v
> v[which(v >= 0)] # nonnegative elements of v
```

9.5 Examples

One of the several merits of R is in the diversity and convenience of its modelling functions. The *formula object*, which provides a shorthand method to describe the exact model to be fit to the data, is important to modelling in R. Modelling functions in R usually require a formula object as an argument. The modeling functions return a *model object* that contains all the information about the fit. Generic R functions such as print, summary, plot, anova will have methods defined for precise object classes to return information that is proper for that type of object.

9.5.1   Linear Regression

One of the most common modelling approaches in statistical learning is linear regression. In R, use the `lm` function to generate these models. The following commands generate the linear regression model and give a summary of it.

```
> lm_model <-lm(y ~ x1 + x2,
data=as.data.frame(cbind(y,x1,x2)))

> summary(lm_model)
```

The vector of coefficients for the model is contained in `lm_model$ coefficients`.

9.5.2   Logistic Regression

There is no need to install an extra package for logistic regression. Using the same notation as above (linear regression), the command is:

```
> glm_mod <-glm(y ~ x1+x2,
family=binomial(link="logit"),
data=as.data.frame(cbind(y,x1,x2)))
```

9.5.3   Prediction

To make predictions, we use the predict function. Just type `?predict.name`, where name is the function corresponding to the algorithm. Normally, the first argument is the variable in which you saved the model, and the second argument is a matrix or data frame of test data. When you call the function, you can just type predict instead of predict.name. For example, if we were to predict for the linear regression model above, and `x1_test` and `x2_test` are vectors containing test data, we can use the command:

```
> predicted_values <-predict(lm_model,
newdata=as.data.frame(cbind(x1_test, x2_test)))
```

9.5.4 k-Nearest Neighbour Classification

Install and load the class package. Let X_train and X_test be matrices of the training and test data, respectively, and labels be a binary vector of class attributes for the training examples. For *k* equal to *K*, the command is:

```
> knn_model <-knn(train=X_train, test=X_test,
cl=as.factor(labels), k=K)
```

Then knn_model is a factor vector of class attributes for the test set.

9.5.5 Naive Bayes

Install and load the e1071 package. Or, go to the https://cran.r-project.org/web/packages/e1071/index.html and get the windows Finding the windows binary file[3] binaries for the package and put it in your library folder within the R installation. Then load the library within R.

The command is:

```
> nB_model <-naiveBayes(y ~ x1 + x2,
data=as.data.frame(cbind(y,x1,x2)))
```

You can also try:

```
library(e1071)

x <- cbind(x_train,y_train)

# Fitting model

fit <-naiveBayes(y_train ~ ., data = x)

summary(fit)

#Predict Output

predicted= predict(fit,x_test)
```

---

[3]http://i.stack.imgur.com/TaweP.jpg

9.5.6 Decision Trees (CART)

Decision trees or recursive partitioning are a simple yet powerful tool in predictive statistics. The idea is to split the co-variable space into many partitions and to fit a constant model of the response variable in each partition. CART (classification and regression trees) is implemented in the rpart package. The command is:

```
> cart_model <-rpart(y ~ x1 + x2,
data=as.data.frame(cbind(y,x1,x2)), method="class")
```

You can use plot.rpart and text.rpart to plot the decision tree. You can also try the following:

```
> install.packages('rpart')

> library(rpart)

> #Train the decision tree

> treemodel <- rpart(Species~., data=iristrain)

> plot(treemodel)

> text(treemodel, use.n=T)

> #Predict using the decision tree

> prediction <- predict(treemodel, newdata=iristest,
type='class')

> #Use contingency table to see how accurate it is

> table(prediction, iristest$Species)
```

prediction	setosa	versicolor	virginica
setosa	10	0	0
versicolor	0	10	3
virginica	0	0	7

9.5.7 k-Means Clustering

There is no need an extra package. If X is the data matrix and m is the number of clusters, then the command is:

```
> kmeans_model <-kmeans(x=X, centers=m)
```

You can also try:

```
library(cluster)

fit <- kmeans(X, 3) # 5 cluster solution
```

Moreover, in k-Means, you not only require the distance function to be defined, but also the mean function to be specified. You also need k (the number of centroids) to be specified.

k-Means is highly scalable with $O(n * k * r)$ where r is the number of rounds, which is a constant depends on the initial choice of centroids. (Fig. 9.1)

```
> km <- kmeans(iris[,1:4], 3)

> plot(iris[,1], iris[,2], col=km$cluster)

> points(km$centers[,c(1,2)], col=1:3, pch=8, cex=2)

> table(km$cluster, iris$Species)
```

```
     setosa versicolor virginica

1       0        46         50

2      33         0          0

3      17         4          0
```

The output figure is,

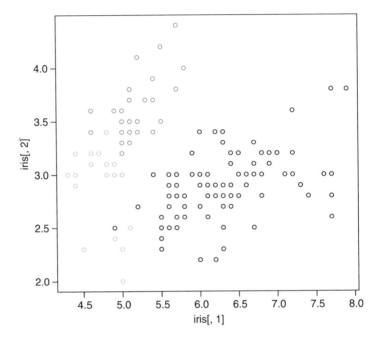

Fig. 9.1 *k*-means plot

9.5.8 *Random Forest*

Random Forest is an important bagging model. It uses multiple models for better performance than just using a single tree model. In addition to selecting n training data out of *N*, at each decision node of the tree, it randomly selects m input features from the total M input features and learns a decision tree from it, and each tree in the forest votes for the output.

```
> install.packages('randomForest')

> library(randomForest)

> #Train 100 trees, random selected attributes

> model <- randomForest(Species~., data=iristrain,
nTree=500)

> #Predict using the forest

> prediction <- predict(model, newdata=iristest,
type='class')

> table(prediction, iristest$Species)

> importance(model)
```

	MeanDecreaseGini
Sepal.Length	7.807602
Sepal.Width	1.677239
Petal.Length	31.145822
Petal.Width	38.617223

9.5.9 Apriori

To run the Apriori algorithm, first install the arules package and load it. Here is an example of how to run the Apriori algorithm using the Mushroom data set.[4] Note that the data set must be a binary incidence matrix; the column names should correspond to the 'items' that make up the 'transactions'. The following commands print out a summary of the results and a list of the generated rules.

---

[4]http://archive.ics.uci.edu/ml/datasets/Mushroom

```
> dataset <-read.csv("C:\\Datasets\\mushroom.csv",
header = TRUE) > mushroom_rules <-
apriori(as.matrix(dataset), parameter = list(supp =
0.8, conf = 0.9)) > summary(mushroom_rules) >
inspect(mushroom_rules)
```

You can adjust the parameter settings as per your preferred support and confidence thresholds.

9.5.10   AdaBoost

There are a number of distinct boosting functions in R. This implementation uses decision trees as base classifiers. Thus, the `rpart` package should be loaded. Also, the boosting function `ada` is in the `ada` package. Let X be the matrix of features, and the labels be a vector of 0–1 class labels. The command is:

```
> boost_model <-ada(x=X, y=labels)
```

9.5.11   Dimensionality Reduction

Dimension reduction signifies the process of converting a set of data having several dimensions into data with smaller dimensions, confirming that it delivers similar information in brief. These techniques are used in solving machine learning problems to obtain better features for a classification or regression task. We can use the `prcomp` function from the stats package to do the Principal Component Analysis (PCA). The PCA is a statistical procedure that transforms and converts a data set into a new data set containing linearly uncorrelated variables, known as principal components. The idea is that the data set is transformed into a set of components where each one attempts to capture as much of the variance (information) in data as possible.

```
library(stats)

pca <- prcomp(train, cor = TRUE)

train_reduced  <- predict(pca,train)

test_reduced  <- predict(pca,test)
```

9.5.12 Support Vector Machine

Support Vector Machines are based on the concept of decision planes that define decision boundaries. A decision plane is one that separates a set of objects having different class memberships. SVMs deliver state-of-the-art performance in real world applications such as text categorization, hand-written character recognition, image classification, bio-sequences analysis, and so on.

```
library("e1071")

# Using IRIS data

head(iris,5)

attach(iris)

# Divide Iris data to x (contains all features) and y
only the classes

x <- subset(iris, select=-Species) y <- Species

# Create SVM model and show summary

svm_model <- svm(Species ~ ., data=iris)
summary(svm_model)

# Run Prediction and measure the execution time

pred <- predict(svm_model1,x)

system.time(pred <- predict(svm_model1,x))

table(pred,y)
```

9.5.13 Artificial Neural Nets

Artificial neural networks are commonly used for classification in data science. They group feature vectors into classes, allowing you to input new data and find out which label fits best. The network is a set of artificial neurons, connected like neurons in the brain. It learns associations by seeing lots of examples of each class, and learning the similarities and differences between them.

```
> library(neuralnet)

> nnet_iristrain <-iristrain

> nnet_iristrain <- cbind(nnet_iristrain,
  iristrain$Species == 'setosa')

> nnet_iristrain <- cbind(nnet_iristrain,
   iristrain$Species == 'versicolor')

> nnet_iristrain <- cbind(nnet_iristrain,
   iristrain$Species == 'virginica')

> names(nnet_iristrain)[6] <- 'setosa'

> names(nnet_iristrain)[7] <- 'versicolor'

> names(nnet_iristrain)[8] <- 'virginica'

> nn <- neuralnet(setosa+versicolor+virginica ~
   Sepal.Length+Sepal.Width+Petal.Length
  +Petal.Width,data=nnet_iristrain,hidden=c(3))

> plot(nn)

> mypredict <- compute(nn, iristest[-5])$net.result

> # Set multiple binary output to categorical output

> maxidx <- function(arr) {
return(which(arr == max(arr)))
  }
```

```
> idx <- apply(mypredict, c(1), maxidx)

> prediction <- c('setosa', 'versicolor',
'virginica')[idx]

> table(prediction, iristest$Species)
```

```
prediction    setosa versicolor virginica

  setosa          10           0          0

  versicolor       0          10          3

  virginica        0           0          7
```

9.6   Visualization in R

R has very flexible built-in graphing capabilities, allowing you to create quality visualizations. A very significant feature of R is its ability to create data visualizations with a couple of lines of code.

The basic plotting command is:

```
plot(x,y) #x and y are the two numbers, vector
variables, or data frame columns to plot
```

This will most often plot an *x–y* scatterplot, but if the x variable is categorical (i.e., a set of names), R will automatically plot a box-and-whisker plot.

We can use the mtcars data set installed with R by default. To plot the engine displacement column disp is on the x axis and mpg on y: (Fig. 9.2)

```
> plot(mtcars$disp, mtcars$mpg)
```

Further, if you would like to label your x and y axes, use the parameters xlab and ylab and for easier reading with the las=1 argument: (Fig. 9.3)

```
> plot(mtcars$disp, mtcars$mpg, xlab="Engine
displacement", ylab="mpg", main="Comparison of MPG &
engine displacement", las=1)
```

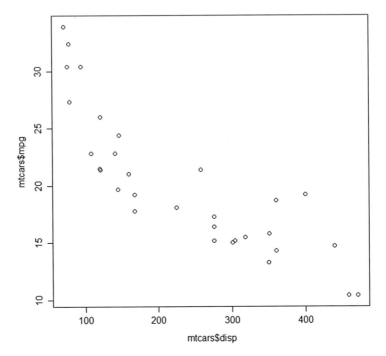

Fig. 9.2 Scatterplot in R

To make a bar graph from the sample BOD data frame included with R, the basic R function is barplot(). So, to plot the demand column from the BOD data set on a bar graph, you can use the command:

```
> barplot(BOD$demand)
```

To label the bars on the x axis, use the names.arg argument and set it to the column you want to use for labels: (Fig. 9.4)

```
> barplot(BOD$demand, main="Bar chart", names.arg =
BOD$Time)
```

There are many more graphics types in R. You can save your R graphics to a file for use outside the R environment. RStudio has an export option in the plots tab of the bottom right window.

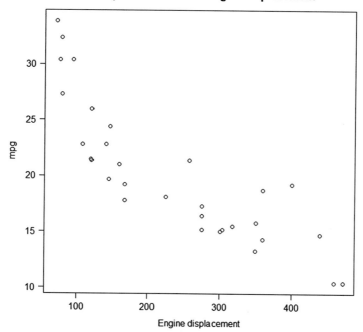

Fig. 9.3 Including a title and axes labels in an R plot

If you are using plain R in Windows, you can also right-click the graphics window to save the file. To save a plot with R commands and not point-and-click, first create a container for your image using a function such as jpeg (), png (), svg () or pdf (). Those functions need to have a file name as one argument and optionally width and height, such as:

```
jpeg("myplot.jpg", width=350, height=420)
```

Other commands include:

```
boxplot(x) #make a box-and-whisker plot

hist(x) #make a histogram

stem(x) #make a stem-and-leaf plot

pie(x) #make a pie diagram
```

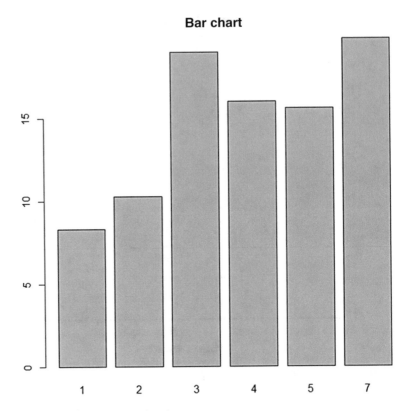

Fig. 9.4 Bar chart with bar plot() function

9.7 Writing Your Own Functions

If R is missing a needed function, write your own. It is very simple to write your own functions in R. Once these are written, they become permanent functions in your version of R, available just as any other function. Custom functions are also useful if you need to create a simple function to work inside `apply()` or `sapply()` commands. Assign the function to a named command just as you would assign data to a variable.

`function.name <- function(x)` `#function_name` is the name of the function, and function(x) tells R it is a function that will have one input variable.

Let us consider the following function that calculates the product of a and b, then finds the mean value, and ultimately takes its square root.

```
function.name<-function(a,b)

{

ab.product<-a*b

ab.avg<-mean(ab.product)

ab.sqrt<-sqrt(ab.avg)

ab.sqrt

}
```

Here's another example of a function named sep that calculates the standard error of an estimate of a proportion. The argument 'n' refers to sample size, and 'X' is the number of 'successes'.

```
sep <- function(X, n){

  p.hat <- X / n     # The proportion of "successes"

  sep <- sqrt(p.hat*(1-p.hat)/(n-1)) # The standard
error of p.hat

  return(sep) # Return the standard error as the
result

  }
```

The function sep will be stored in your R workspace so you only need to paste it once. If you save your workspace when you exit R, it will remain there when you start up again; otherwise you will need to paste it in again.

To use the function on some data, say n = 15 and X = 5, enter

```
> sep(X = 5, n = 15) # yields the standard error

[1] 0.1259882

> sep(5,15)          # ok if X and n are given in
proper order

[1] 0.1259882
```

If you are writing functions, it's usually a good idea to not let your functions run for pages and pages. Typically, the purpose of a function is to execute one activity or idea. If your function is doing lots of things, it probably needs to be broken into multiple functions.

Note that the variable names within the function do not need to correspond to the actual variables you want to evaluate; so to run the function on data, just type.

9.8   Open-Source R on Hadoop

Data engineers continue to explore different approaches to leverage the distributed computation potential of MapReduce and the unlimited storage capacity of HDFS in ways that can be exploited by R. We list in the following lines some packages that address the scalability issue of R to different levels.

RHive This open-source R package allows R users to run HIVE queries and offers custom HIVE UDFs to embed R code into the HIVE queries. Although there is no inherent 'transparency' involved, R users need to know HQL to explicitly write the queries and pass them into the RHive functions. You need to install Rserver and RJava packages to make it work. The performance impact of these additional packages is unclear.

RHadoop/RMR2 This is one of the popular R packages for Hadoop. It has a simple interface for running MapReduce jobs using Hadoop streaming. It also has a set of interfaces to interact with HDFS files and Hbase tables. Alongside its companion packages rhdfs and rhbase (for working with HDFS and HBASE datastores, respectively, in R) the RMR package provides a way for data analysts to access massive, fault tolerant parallelism without needing to master distributed programming.

RHIPE Rhipe is a software package that allows the R user to create MapReduce jobs which work entirely within the R environment using R expressions. This integration with R is a transformative change to MapReduce; it allows an analyst to promptly specify Maps and Reduces using the full power, flexibility, and expressiveness of the R interpreted language. There is a custom jar file that invokes the R processes in each cluster and uses protocol buffers for data serialization/de-serialization. However, there is no transparency layer to perform ad-hoc analytics and data preparation.

Revolution R It is a commercial R offering support for R integration on Hadoop distributed systems. Revolution R ensures to provide improved performance, functionality, and usability for R on Hadoop. To provide deep analytics like R, Revolution R makes use of the company's ScaleR library, which is a set of statistical analysis algorithms developed precisely for enterprise-scale big data collections.

References

An Introduction to R: https://cran.r-project.org/doc/manuals/R-intro.html

The R Journal (http://journal.r-project.org/index.html): The refereed journal with research articles introducing new R packages, updates on news and events, and textbook reviews.

www.r-project.org: The R Project home page. You will find official documentation, links to download all versions of R, information about the R contributors and foundation, and much more.

Appendices

Appendix I: Tools for Data Science

In this appendix, we present some useful tools for prominent aspects of data science; namely, data mining, machine learning, soft computing, predictive analytics, and business intelligence. We have discussed R, in Chap. 9, which is widely used and praised for its rich set of industry-standard algorithms, publication-quality plotting, rapid prototyping, and native functional programming.

BigML[1]

BigML is a cloud based machine learning platform with an easy to use graphical interface. It also provides simple mechanisms to incorporate predictive models into production applications through its REST API. The platform combines supervised learning (to build predictive models), unsupervised learning (to understand behavior), anomaly detection (used in fraud detection), data visualization tools (scatter-plots and Sunburst diagrams) and many mechanisms for exploring data. Resulting models can be exported into Java, Python NOde.js and Ruby code as well as Tableau or Excel format.

---

[1]https://bigml.com/

Python[2]

Python is a popular general-purpose, high-level programming language with intuitive and aesthetic syntax, but a runtime that can be slow compared to compiled languages like C++ and Java The creation in 2005 of NumPy, a library for very fast matrix numeric computation, spurred the use of Python within machine learning communities who might have previously used MATLAB or C.

Natural Language Toolkit

The Natural Language Toolkit (NLTK[3]) was developed in conjunction with a computational linguistics course at the University of Pennsylvania in 2001. It was designed with three pedagogical applications in mind: assignments, demonstrations, and projects. It is a set of Python libraries that makes it easy to carry out common tasks involved in processing natural language, such as tokenization and stemming, creating frequency lists and concordances, part-of-speech tagging, parsing, named entity recognition, semantic analysis, and text classification.

DataWrangler[4]

DataWrangler is an interactive tool for data cleaning and transformation from Stanford University's Visualization Group. Wrangler allows interactive transformation of messy, real world data into the data tables analysis tools expect.

OpenRefine[5]

OpenRefine is formerly called Google Refine, and is used for working with messy data: cleaning it, transforming it from one format into another, and extending it with web services and external data.

---

[2]https://www.python.org/

[3]http://www.nltk.org/

[4]http://vis.stanford.edu/wrangler/

[5]http://code.google.com/p/google-refine/

Datawrapper

Datawrapper was created by journalism organizations from Europe, designed to make data visualization easy for news institutes. Based on a web-based GUI (graphics user interface), it promises to let you create a graph in just four steps. Datawrapper is fully open source, and you can download it from their GitHub page and host it yourself. It is also available as a cloud-hosted, paid service on their website.

Orange

Orange is an open source data mining, visualization environment, analytics, and scripting environment. The Orange environment, paired with its array of widgets, supports most common data science tasks. Orange supports scripting in Python as well as the ability to write extension in C++. However, support for big data processing is not available.

RapidMiner

RapidMiner, formerly Yale, has morphed into a licensed software product as opposed to being open source. RapidMiner has the ability to perform process control (i.e., loops), connect to a repository, import and export data, and perform data transformation, modelling (i.e., classification and regression), and evaluation. RapidMiner is full featured with the ability to visually program control structures in the process flows. Modelling covers the key methods such as decision trees, neural networks, logistic and linear regression, support vector machines, naïve Bayes, and clustering. In some cases (say, k-means clustering) multiple algorithms are implemented, leaving the data scientist with options. RapidMiner's Radoop is especially designed for big data processing, but it is not available in the free edition.

Tanagra

Tanagra claims to be an open source environment for teaching and research, and is the successor to the SPINA software. Capabilities include Data source (reading of data), Visualization, Descriptive statistics, Instance selection, Feature selection, Feature construction, Regression, Factorial analysis, Clustering, Supervised learning, Meta-Spv learning (i.e., bagging and boosting), Learning assessment, and Association rules.

Weka[6]

Weka, or the Waikato Environment for Knowledge Analysis, is licensed under the GNU general public license. Weka stems from the University of Waikato, is Java based, and is a collection of packages for machine learning. For big data processing, Weka has its own packages for map reduce programming to maintain independence over the platform, but also provides wrappers for Hadoop.

KNIME

KNIME is the Konstant Information Miner, which had its beginnings at the University of Konstanz, and has since developed into a full-scale data science tool. Big data processing is not included in the free version, but may be purchased as the KNIME Big Data Extension.

Apache Mahout[7]

In order to gain a deep understanding of machine learning (such as k-Means, Naïve Bayes, etc.) and statistics, and how these algorithms can be implemented efficiently for large data sets, the first place to look is *Mahout*, which implements many of these algorithms over Hadoop.

Hive[8]

Typically, SQL programmers have been familiar with data for many years and understand well how to use data to gain business insights. *Hive* gives you access to large data sets on Hadoop with familiar SQL primitives. It is an easy first step into the world of big data. From a querying perspective, using Apache Hive provides a familiar interface to data held in a Hadoop cluster, and is a great way to get started. Apache Hive is data warehouse infrastructure built on top of Apache Hadoop for providing data summarization, ad hoc query, and analysis of large data sets.

---

[6]http://www.cs.waikato.ac.nz/ml/weka/

[7]http://mahout.apache.org/

[8]http://hive.apache.org/

Scikit-Learn[9]

Scikit-learn provides a range of supervised and unsupervised learning algorithms via a consistent interface in Python. It is distributed under many Linux distributions, encouraging academic and commercial use. This package focuses on bringing machine learning to non-specialists using a general-purpose, high-level language.

D3.js[10]

D3.js is a comprehensive JavaScript library that makes it pretty simple to develop rich, interactive, web- and production-ready visualizations. D3 helps you bring data to life using HTML, SVG, and CSS. D3's emphasis on web standards gives you the full capabilities of modern browsers without tying yourself to a proprietary framework, combining powerful visualization components and a data-driven approach to DOM manipulation. Tools for web-based data visualization tend to evolve and specialize extremely fast, but D3 has become something of a standard, and is frequently used alongside and within other new libraries.

Pandas[11]

Pandas is an open source library providing high-performance, easy-to-use data structures and data analysis tools for the Python programming language. It provides fast, flexible, and expressive data structures designed to make working with structured (tabular, multidimensional, potentially heterogeneous) and time series data both easy and intuitive. It aims to be the fundamental high-level building block for doing practical, real world data analysis in Python.

Tableau Public[12]

Tableau Public is one of the most popular visualization tools and supports a wide variety of charts, graphs, maps, and other graphics. It is a free tool, and the charts you make with it can be easily embedded in any web page. However, Tableau Public

---

[9]http://scikit-learn.org/stable/

[10]http://d3js.org/

[11]http://pandas.pydata.org/

[12]https://public.tableau.com/

does not have a central portal or a place to browse data. Rather, it is focused on letting you explore data and stitch modules together on your desktop and then embed your findings on a website or blog.

Exhibit[13]

Exhibit (part of the SIMILE Project) is a lightweight, structured data publishing framework that allows developers to create web pages with support for sorting, filtering, and rich visualizations. Focused on semantic web-type problems, Exhibit can be implemented by writing rich data out to HTML then configuring some CSS and JavaScript code. Exhibit allows user to effortlessly create web pages with advanced text search and filtering functionalities, with interactive maps, timelines, and other visualizations.

Gephi[14]

Gephi is an easy access and powerful network analysis tool. It is also known as the 'Photoshop for networks'. Visualizing your data as a network of interconnections helps reveal patterns and behaviour, analyse interconnections, and discover entries that appear to be anomalies. It uses a 3D render engine, the same technology that many video games use, to create visually amazing graphs. Gephi can be used to explore, analyse, spatialize, filter, cluster, manipulate, and export all types of networks.

NodeXL[15]

NodeXL is a powerful and easy-to-use interactive network visualization and analysis tool that leverages the widely available MS Excel application as the platform for representing generic graph data, performing advanced network analysis and visual exploration of networks. The tool supports multiple social network data providers that import graph data (nodes and edge lists) into Excel spreadsheets.

---

[13] http://simile-widgets.org/exhibit/

[14] https://gephi.org/

[15] https://nodexl.codeplex.com/

Leaflet[16]

Mobile readiness is the key to high traffic and good conversion rates. Leaflet is a lightweight, mobile friendly JavaScript library to help you create interactive maps. Leaflet is designed with simplicity, performance, and usability in mind. It works across all major desktop and mobile platforms out of the box, taking advantage of HTML5 and CSS3 on modern browsers while still being accessible on older ones. It can be extended with a huge number of plugins, has a beautiful, easy to use, and well-documented API and a simple, readable source code that is a joy to contribute to.

Classias[17]

Classias is a collection of machine-learning algorithms for classification. The source code of the core implementation is well structured and reusable; it provides components such as loss functions, instance data-structures, feature generators, online training algorithms, batch training algorithms, performance counters, and parameter exchangers. It is very easy to write an application on top of these components.

Appendix II: Tools for Computational Intelligence

NeuroXL[18]

NeuroXL Classifier is a fast, powerful and easy-to-use neural network software tool for classifying data in Microsoft Excel. NeuroXL Predictor is a neural network forecasting tool that quickly and accurately solves forecasting and estimation problems in Microsoft Excel. In NeuroXL products, all the science is hidden behind the curtain, and only what is really important for the user is left: what they have and what they get. NeuroXL combines the power of neural networks and Microsoft Excel, and it is easy for almost anyone to use. With NeuroXL, you can do all the hard work in a few seconds.

---

[16]http://leafletjs.com/

[17]http://www.chokkan.org/software/classias/

[18]http://www.neuroxl.com/

Plug&Score[19]

Plug&Score is credit scoring software based on logistic regression for scorecard development, validation, and monitoring. A scorecard can be deployed with one click to Plug&Score Engine for automated processing of real-time scoring requests.

Multiple Back-Propagation (MBP)[20]

Multiple Back-Propagation is a free application for training neural networks with the Back-Propagation and the Multiple Back-Propagation algorithms, with the ability to generate C code for the trained networks.

A.I. Solver Studio[21]

A.I. Solver Studio is a unique pattern recognition application that deals with finding optimal solutions to classification problems, and uses several powerful and proven artificial intelligence techniques, including neural networks, genetic programming, and genetic algorithms. No special knowledge is required of users, as A.I. Solver Studio manages all the complexities of the problem solving internally. This leaves users free to concentrate on formulating their problems of interest.

The MathWorks – Neural Network Toolbox[22]

This toolbox provides a complete set of functions and a graphical user interface for the design, implementation, visualization, and simulation of neural networks. It supports the most commonly used supervised and unsupervised network architectures and a comprehensive set of training and learning functions.

---

[19]http://www.plug-n-score.com

[20]http://dit.ipg.pt/MBP/

[21]http://www.practical-ai-solutions.com/AISolverStudio.aspx

[22]http://www.mathworks.com/products/neuralnet/

Visual Numerics Java Numerical Library[23]

The JMSL Library is the broadest collection of mathematical, statistical, financial, data mining, and charting classes available in 100 % Java. The JMSL Library now includes neural network technology that adds to the broad selection of existing data mining, modelling, and prediction technologies available across the IMSL family of products.

Stuttgart Neural Network Simulator[24]

SNNS (Stuttgart Neural Network Simulator) is a software simulator for neural networks on Unix workstations developed at the Institute for Parallel and Distributed High Performance Systems (IPVR) at the University of Stuttgart. The goal of the SNNS project is to create an efficient and flexible simulation environment for research on and application of neural nets.

FANN (Fast Artificial Neural Network Library)[25]

FANN Library is written in in ANSI C. The library implements multilayer feedforward ANNs, up to 150 times faster than other libraries. FANN supports execution in fixed point, for fast execution on systems such as the iPAQ.

NeuroIntelligence – Alyuda Research[26]

NeuroIntelligence is neural network software for experts designed for intelligent support in applying neural networks to solve real world forecasting, classification, and function approximation problems.

---

[23] http://www.vni.com/products/imsl/jmsl/jmsl.html

[24] http://www-ra.informatik.uni-tuebingen.de/SNNS/

[25] http://sourceforge.net/projects/fann

[26] http://www.alyuda.com/neural-network-software.htm

EasyNN-Plus[27]

EasyNN-plus is a neural network software system for Microsoft Windows based on EasyNN that can generate multi-layer neural networks from imported files or grids with minimal user intervention. Neural networks produced by EasyNN-plus can be used for data analysis, prediction, forecasting, classification, and time series projection.

NeuroDimension – Neural Network Software[28]

NeuroSolutions is a powerful and flexible neural network modelling software, and is the perfect tool for solving data modelling problems.

BrainMaker – California Scientific Software[29]

BrainMaker Neural Network Software lets you use your computer for business and marketing forecasting, and for stock, bond, commodity, and futures prediction, pattern recognition, and medical diagnosis.

Classification & Prediction Tools in Excel[30]

A set of Open Source tools for data mining and prediction within Excel, using Neural Networks. NNpred, NNclass, and Ctree are a set of three tools in Excel for building prediction and classification models. They are free and open source. These tools are good for small- to medium-sized data sets. The aim here is to provide easy-to-use tools for beginners in the areas of neural network, prediction, and classification problems. Users can put their own data sets in the tools, play with various parameters, and study the effect of final result. Such tools will be useful for the classroom teaching setting as well as for quickly building small prototypes.

---

[27] http://www.easynn.com/

[28] http://www.neurodimension.com/

[29] http://www.calsci.com/BrainIndex.html

[30] http://www.sites.google.com/site/sayhello2angshu/dminexcel

SIMBRAIN[31]

SIMBRAIN is a freeware tool for building, running, and analysing neural networks. SIMBRAIN aims to be as visual and easy to use as possible. Also unique are its integrated *world component* and its representation of the network's activation space. SIMBRAIN is written in Java and runs on Windows, Mac OS X, and Linux.

DELVE[32]

DELVE is a standard environment for evaluating the performance of learning methods. It includes a number of data sets and an archive of learning methods.

SkyMind[33]

Skymind is the commercial support arm of the open-source framework Deeplearning4j, bringing the power of deep learning to enterprise on Hadoop and Spark.

Prediction.io[34]

Prediction.io is an open source, machine learning server for developers and data scientists to create predictive engines for production environments, with zero-downtime training and deployment.

Parallel Genetic Algorithm Library (pgapack[35]*)*

PGAPack is a general-purpose, data-structure-neutral, parallel genetic algorithm library developed at Argonne National Laboratory.

---

[31] http://simbrain.sourceforge.net/

[32] http://www.cs.toronto.edu/~delve/

[33] http://www.skymind.io/

[34] https://prediction.io/

[35] https://code.google.com/p/pgapack/

Parallel PIKAIA

If you wish to use the genetic algorithm-based FORTRAN-77 optimization subroutine PIKAIA,[36] and the modelling fitness-function that you want to maximize is computationally intensive, then Parallel PIKAIA is good option.

Evolving Objects (EO): An Evolutionary Computation Framework[37]

EO is *a template-based, ANSI-C++ evolutionary computation library* that helps you to write your own stochastic optimization algorithm. These are stochastic algorithms, because they iteratively use random processes. The vast majority of these methods are used to solve optimization problems, and may also be called 'metaheuristics'.

---

[36]http://www.hao.ucar.edu/public/research/si/pikaia/pikaia.html

[37]http://eodev.sourceforge.net/